Stimulus-Secretion Coupling in Chromaffin Cells

Volume II

Editors

Kurt Rosenheck, Ph.D.
Associate Professor
Department of Membrane Research
The Weizmann Institute of Science
Rehovot, Israel

Peter I. Lelkes, Ph.D.
Laboratory of Cell Biology and Genetics
National Institutes of Health
Bethesda, Maryland

CRC Press, Inc.
Boca Raton, Florida

Library of Congress Cataloging-in-Publication Data

Stimulus-secretion coupling in chromaffin cells.
  Includes bibliographies and index.
  1. Chromaffin cells. 2. Secretion. I. Rosenheck,
Kurt. II. Lelkes, Peter I. [DNLM: 1. Adrenal Medulla—
secretion. 2. Chromaffin System—secretion.
WK 725 S859]
QP188.A33S75   1987   599'.0142   87-6378
ISBN 0-8493-6534-1 (set)
ISBN 0-8493-6536-8 (v. 1)
ISBN 0-8493-6537-6 (v. 2)

International Standard Book Number 0-8493-6536-8(v. 1)
International Standard Book Number 0-8493-6537-6(v. 2)
International Standard Book Number 0-8493-6534-1(set)

Library of Congress Card Number 87-6378
Printed in the United States

INTRODUCTION

The adrenal chromaffin cell occupies a unique place in modern cell and neurobiology. Originating from the neural crest, chromaffin cells are believed to be modified postganglionic sympathetic neurons. Thus, over the past two decades, they have served as one of the most readily available model systems in studying the mechanism(s) of neurotransmitter release. The pioneering experiments of W. W. Douglas and his colleagues in adrenal chromaffin cells have firmly established the now widely accepted concept of calcium-dependent stimulus-secretion coupling as a universal mechanism for the exocytotic secretion of a variety of neurotransmitters and hormones, packaged in intracellular storage organelles of many different cell types.

Over the years, a number of excellent reviews on chromaffin cells have been published, most of them dealing with general aspects of chromaffin cell biology. However, during our own involvement in the study of the mechanism of catecholamine secretion, we felt the need for a more specialized treatise on what we believe the be the central *raison de etre* of chromaffin cells: catecholamines and neuropeptides, synthesized and stored in the cells, are released in a strictly controlled fashion and upon special physiological demand only. So what then are those mechanisms that translate the extracellular signals into the cellular responses, culminating in the secretion of storage products? What do we really know about the intracellular mechanism linking cell activation to secretion?

In order to address these questions in as broad a fashion as possible, we have gathered for these volumes authoritative contributions of leading experts on the adrenal chromaffin cell, and in particular, on the mechanism(s) of stimulus-secretion coupling. As it is unavoidable in multiauthored monographs like this, some of the contributions contain several seemingly redundant treatises of similar issues; however, we deliberately invited such complementary chapters, since every author is presenting his individual concept of a very complex issue.

As is evident from studying the various chapters, we are still far from being able to draw a unified picture of what might be going on during the processes of stimulus-secretion coupling in the adrenal chromaffin cell. On the contrary, the more detailed our information becomes about the intracellular events following stimulation of the cells and leading up to the exocytotic secretion, the more emerges our obvious lack in understanding of the finer tunings of the cell biology of the chromaffin cells. At present, exciting new developments are rapidly changing some of our basic understanding about, for example, membrane biophysics of ion channels, intracellular second and third messengers, the temporal and spatial role of calcium homeostasis, or the role of cytosolic proteins in mediating exocytotic membrane fusion, to name but a few of the unresolved issues. Furthermore, with more details emerging on the biological, biophysical, and molecular biological patterns of stimulus-secretion coupling in chromaffin cells, the more the individuality of this particular cell will become evident.

In editing this book, we therefore attempted to provide, for the first time, an in-depth resume of several aspects of our current (as of 1986) understanding of stimulus-secretion coupling. The authors were asked not only to provide a state-of-the-art review in their respective fields of interest, but also to look ahead and try to address unresolved issues and those relevant questions to be tackled in the near future. Thus, this monograph, centered around a key function in the cellular biology of adrenal chromaffin cells, combines solid evidence on what is presently known as well as more speculative individual assessments as to future developments. Concomitantly, however, this book is published with the cautionary notion on extrapolating our current knowledge about chromaffin cells to apparently similar mechanisms governing stimulus-secretion coupling in other secretory cells or to different endocrine organs.

For technical reasons the 15 chapters have been arranged in two volumes. Nevertheless, these two volumes are intended as a contiguous, single monograph on the multifaceted issue termed "stimulus-secretion coupling".

In the first chapter, S. W. Carmichael presents a detailed description of the anatomical morphology of the adrenal medulla. J. H. Phillips deals in two chapters with an extensive discussion of chromaffin granules: beyond the biogenesis of these storage organelles, their structure, and their dynamics, the author discusses the fate of chromaffin granules during the perpetual cycles of exocytosis and endocytosis.

More recently, a number of bioactive peptides, in particular enkephalins, were found to be localized in chromaffin granules and co-released together with the catecholamines; these exciting new findings are summarized by C. D. Unsworth and O. H. Viveros in Chapter 4.

Calcium is generally believed to be of central importance for a number of biochemical processes linking cell stimulation to exocytotic secretion. One of the unresolved issues relates to the question of calcium buffering and calcium homeostasis in chromaffin cells and the possible role of the granules in these processes. In Chapter 5, Gratzl discusses possible pathways for the uptake of calcium into chromaffin granules and the relevance of Na^+/Ca^{2+} exchange across the granules for intracellular calcium homeostasis.

Specific calcium-regulating and -regulated cytosolic proteins, which are believed to be involved in intracellular signal transduction, are discussed in the next group of chapters. As described in Chapter 6, Trifaró and Kenigsberg used classical pharmacological approaches as well the fusion of antibody loaded erythrocyte ghosts with cultured chromaffin cells to probe the central role for calmodulin in stimulus-secretion coupling. The relevance of cytoskeletal proteins in intracellular signal transduction, especially the structural and functional role of a spectrin-related, actin-associated protein α-fodrin is reviewed by Aunis, Perrin, and Langley in Chapter 7. In recent years, a number of cytosolic proteins, such as the phospholipases, proteinkinases, etc., have been proposed to mediate calcium action during exocytosis. In Chapter 8, Pollard and his colleagues primarily discuss the family of synexins and immunologically related proteins, which seem to regulate membrane contact and fusion in a calcium-dependent fashion. P. I. Lelkes describes (Chapter 9) how liposomal vectors can be employed to introduce bioactive (macro) molecules, such as cytoskeletal proteins, etc., into intact chromaffin cells to study their involvement in the cascades of stimulus-secretion coupling.

Recognition of the stimulus at the plasma membrane is the primary event in activating a cellular response. Yet, as summarized by Rosenheck in Chapter 10, our present knowledge of the functional biochemistry at the plasma membrane level is quite limited, presumably due to too scarce a usage of isolated chromaffin cell plasma membrane preparations.

Adrenal chromaffin cells are activated via nicotinic and/or muscarinic receptors, depending on the species and probably also on physiological idiosyncrasies. The importance of muscarinic stimulation has recently been emphasized due to the clear linkage of muscarinic activation to the phosphoinositide metabolism in a number of eukaryotic cells. In Chapter 11, Allan Schneider discusses muscarinic receptor mechanisms in adrenal chromaffin cells.

Cell activation comprises transmembranal ion fluxes. In intact cells, the various ion channels of the plasma membrane have been characterized and the ion fluxes pertaining to the intracellular signal transduction have been widely studied. In Chapter 12, Kirshner presents an overview of the more classical biochemistry and electrophysiology of calcium and sodium channels in intact chromaffin cells. However, over the past few years, electrophysiology has been revolutionized by analyzing single channels using the patch-clamp technique. Thus, Kidokoro summarizes in Chapter 13 these more recent developments in membrane biophysics of chromaffin cells.

Inhibitory modulations of the secretory response, often termed desensitization, are discussed in the final two chapters. In Chapter 14, Garcia and his co-workers review the evidence for the pivotal role of calcium channel activation for the onset of stimulus-secretion coupling and the inactivation of the secretory response by sustained elevated intracellular calcium concentrations. Finally, in Chapter 15, Bruce Livett summarizes our current understanding of various parameters, in particular the role of neuropeptides, which in vitro and in vivo are involved in the modulation of the secretory response in adrenal chromaffin cells.

During the entire process of compiling and editing this monograph, we were encouraged by Marsha Baker, the senior editor in charge of the Uniscience Series at CRC Press, and guided by the helpful advice and patience of our coordinating editor, Anita Hetzler. We especially appreciate the enduring understanding and support from our families, who for the many years of our close collaboration have tolerated our time-consuming fascination with the chromaffin cells more or less patiently. We therefore dedicate this book to our families, especially to our wives, Alma Rosenheck and Iris Lelkes.

Rehovoth and Bethesda, December 1986

Kurt Rosenheck
Peter I. Lelkes

THE EDITORS

Kurt Rosenheck received his Ph.D. in Physical Chemistry from the Hebrew University of Jerusalem in 1959 for work on polyelectrolytes, under the direction of Aharon Katzir-Katchalsky. He then was a research assistant in the Polymer Department of the Weizmann Institute of Science, Rehovot, Israel, and subsequently worked as a research fellow with Paul Doty in the Chemistry Department of Harvard University.

He spent 1969 as a resident scientist in the Neuroscience Research Program of the Massachusetts Institute of Technology, Cambridge, directed by Francis O. Schmitt. In 1975 he was a visiting investigator in the Endocrinology Section, directed by Martin Sonenberg, of the New York Sloan-Kettering Institute for Cancer Research. In 1977 he was appointed Associate Professor at the Weizmann Institute's Department of Membrane Research. During the years 1984 and 1986 he held Visiting Professorships at the Universities of Constance and Bielefeld, in West Germany, as well as at the University of Berne in Switzerland.

Starting out as a polymer spectroscopist, he later applied ultraviolet light spectroscopic techniques to the study of protein conformation and lipid-protein interactions in biological membranes. This led to his interest in the membrane-linked events occurring during the stimulus-secretion response, a field he has been active in for the last 10 years.

Peter I. Lelkes, born in 1949 in Budapest, Hungary, received his training in physics, biochemistry, cell biology, and membrane biophysics at the Technical University in Aachen, West Germany. In 1977 he joined the Department of Membrane Research at the Weizmann Institute of Science in Rehovoth, Israel, as a postdoctoral fellow to work on physicochemical aspects of protein-lipid interactions of biological membranes. Subsequently, as a staff scientist in the same department, he focused his interest on the mechanism of membrane fusion of biological membranes, and in particular, on the involvement of cytosolic proteins in the stimulus-secretion coupling in chromaffin cells and other secretory systems.

In 1983, he went to the National Institutes of Health, Bethesda, Md., to further study the pathways of intracellular signal transduction of secretory cells. Working in the National Institute of Diabetes, Digestive, and Kidney Diseases, he currently holds the position of Visiting Scientist in the Laboratory of Cell Biology and Genetics, and continues to study cell biological aspects of the activation of adrenal medullary chromaffin and other endocrine cells.

CONTRIBUTORS

Alexander L. Burns, Ph.D.
Special Expert
Laboratory for Cell Biology and Genetics
NIH/NIDDK
Bethesda, Maryland

Valentin Ceña, M.D., Ph.D.
Professor Titular
Section of Neuroendocrinology
National Institutes of Health
Bethesda, Maryland

Antonio G. Garcia, M.D., Ph.D.
Professor and Chairman
Department of Neurochemistry
University of Alicante
Alicante, Spain

Yoshiaki Kidokoro, Ph.D.
Professor
Department of Physiology
Jerry Lewis Center
UCLA School of Medicine
Los Angeles, California

Norman Kirshner, Ph.D.
Professor and Chairman
Department of Pharmacology
Duke University Medical Center
Durham, North Carolina

M. G. Ladona, M.D.
Department of Neurochemistry
University of Alicante
Alicante, Spain

Peter I. Lelkes, Ph.D.
Visiting Scientist
Laboratory for Cell Biology and Genetics
NIH/NIDDK
Bethesda, Maryland

Bruce G. Livett, Ph.D.
Reader in Biochemistry
Department of Biochemistry
University of Melbourne
Melbourne, Australia

Carmen Montiel, Ph.D.
Professor Titular
Department of Pharmacology
Facultad de Medicine Autonoma
Madrid, Spain

Kyoji Morita
Associate Professor
Laboratory for Cell Biology and Genetics
NIH/NIDDK
Bethesda, Maryland

Harvey B. Pollard, M.D., Ph.D.
Chief
Laboratory for Cell Biology and Genetics
NIH/NIDDK
Bethesda, Maryland

Eduardo Rojas, Ph.D.
Visiting Professor
Laboratory for Cell Biology and Genetics
NIH/NIDDK
Bethesda, Maryland

Kurt Rosenheck, Ph.D.
Associate Professor
Department of Membrane Research
Weizmann Institute of Science
Rehovot, Israel

F. Sala, M.D., Ph.D.,
Department of Neurochemistry
University of Alicante
Alicante, Spain

Allan S. Schneider, Ph.D.
Associate Professor
Department of Pharmacology and
  Toxicology
Albany Medical College
Albany, New York

Andres Stutzin, M.D.
Visiting Fellow
Laboratory for Cell Biology and Genetics
NIH/NIDDK
Bethesda, Maryland

TABLE OF CONTENTS

Volume I

Volume II

Chapter 8

CYTOSOLIC PROTEINS AS INTRACELLULAR MEDIATORS OF CALCIUM ACTION DURING EXOCYTOSIS

Harvey B. Pollard, Alexander L. Burns, Andres Stutzin, Eduardo Rojas, Peter I. Lelkes, and Kyoji Morita

TABLE OF CONTENTS

I. INTRODUCTION

It is now beyond argument that calcium plays a critical role in the regulation of secretion from chromaffin cells. However, exactly what this role entails, or whether there are a multiplicity of such roles remains a mystery. Calcium enters the cytosol upon stimulation and then provokes a cascade of events leading to secretion of chromaffin granule contents, but not the granule membrane. This is "exocytosis". However, even such questions as how calcium enters, from which sources, how much is free, and over what time course calcium remains free are, as yet, unanswered with certainty.

Going beyond the quantitative problems of calcium concentration, the consequences of calcium presence seem to be movement of granules to the plasma membrane, contact with the membrane, and eventual fusion. It was once thought that calcium acted directly on the plasma membrane and granule membrane to induce fusion. Presently there is really no compelling experimental reason to support such a contention. On the other hand, there is a number of specific proteins which could mediate calcium action in these various events. Recent candidates include calmodulin, actin, cytoskeletal proteins, synexin, protein kinase C, phospholipase C, metabolic products of phosphatidylinositol and inositol phosphates, metalloendoproteases, and other as yet unspecified fusion factors.

It is our intention in this chapter to focus attention on cytosolic activities possibly involved in calcium signaling and on two specific proteins that interact with chromaffin granule membranes in a calcium-dependent manner, actin and synexin. It is possible that calcium-dependent interactions with actin model events occurring during granule movement through the cytosol, while interactions with synexin may model membrane contact and fusion events occurring during exocytosic secretion.

II. CALCIUM SIGNAL PROCESSES

A. Phospholipase C

Phospholipase C is becoming increasingly appreciated as an important intermediary in a growing number of secretory cell systems.[2] Phospholipase C cleaves phosphatidyl inositol to liberate inositolphosphate (IP) or more phosphorylated derivatives such as inositolbisphosphate (IP_2) or inositoltrisphosphate (IP_3). In the case of skeletal muscle[3] and other systems the IP_3 is believed to mediate release of calcium from the sarcoplasmic reticulum, or, in the case of nonmuscle cells, the endoplasmic reticulum.

In the case of the chromaffin cell, the possible involvement of IP_3 with secretion has been somewhat mysterious. By analogy with other secretory systems one might expect that physiologic secretagogues might evoke IP_3 synthesis and coincident increases in cytosolic Ca^{2+} concentration from internal stores. Perversely, however, nicotine, the specific cholinergic secretagogue, apparently has no effect on phosphatidylinositol metabolism, whereas muscarine, having no secretagogue activity, does indeed have this effect.[4] Until recently, this fact has lurked uncompromisingly in the back of many otherwise optimistic reviews about the likely generality of phospholipase-C activation being coupled to calcium mobilization and secretion.

The solution to this problem has come from studies on the real-time kinetics of ATP release from chromaffin cells[5] using a newly available, highly purified luciferase to detect quantitatively released ATP over lengthy time periods. We noted that muscarine was actually able to potentiate nicotine-induced ATP release from chromaffin cells if the two agonists were added within at least 1 min of one another. Interestingly, the potentiation was barely observed if the two compounds were added simultaneously. In their subsequent work, Forsberg et al.[6] observed the same phenomena for catecholamine release and noted that IP_3 rose quickly during the first 15 sec after muscarinic stimulation. Others[7] had shown that the calcium concentrations measured by quin-2 rises and falls slightly over a 2-min period when chromaffin cells are exposed to muscarine. Indeed, muscarine is known to cause calcium efflux from chromaffin cells.[8] Thus, the window of opportunity for muscarine to potentiate nicotine-induced release would seem to be an IP_3-mediated rise and fall in intracellular calcium concentration.

To further substantiate this concept, Forsberg et al.[6] noted that a slight depolarization with 10 mM KCl could likewise potentiate nicotine-induced release of both ATP and catecholamines. The only common consequence of muscarine and KCl could, on the basis of present knowledge, only be an elevation in cytosolic calcium concentration.

B. Protein Kinase C

As a further consequence of phospholipase-C activity diacylglyceride is also produced. Diacylglyceride and calcium, together with phosphatidyl serine, activate protein kinase C, and in a variety of systems activation of protein kinase C has also been implicated with activation of secretory processes.[9]

The evidence for protein kinase C being involved in regulation of secretion from chromaffin cells is only indirect, but nonetheless remains encouraging. The tumor-promoting phorbol ester TPA is an effective analog of diacylglycerol for activation of protein kinase C, and this same compound has been found to decrease the calcium concentration dependence of catecholamine release from high voltage permeabilized chromaffin cells.[10] However, phorbol esters had no effect on secretions from intact cells. Morita et al.[11] did find that phorbol ester could potentiate A23187-induced catecholamine secretion, but the ester had no effect on secretion induced by carbachol or high potassium. Recent results from Brocklehurst and Pollard[12] also showed that the phorbol ester TPA could enhance calcium-induced catecholamine release from digitonin-permeabilized cells, and also raise the apparent calcium sensitivity of the process. In this case, high voltage-permeabilized cells and digitonin-permeabilized cells behaved similarly.

These results might indicate that protein kinase C could be involved in modulating secretion induced by elevation in bulk cytosolic calcium concentration, but not secretion induced by secretagogues acting on plasma membrane receptors. The concept of mechanistic significance being accorded these two types of calcium compartments is also found in other chapters of this book. It is becoming increasingly apparent that the mere presence of elevated calcium concentration in the cytosol is actually necessary, but not sufficient, to induce release from chromaffin cells.

Protein kinase C has been extensively studied, but seldom purified from biological tissue. Brocklehurst et al.[13] characterized protein kinase C in crude extracts from bovine adrenal medulla with respect to calcium, diolein, and phorbol ester concentration dependencies. More recent studies by Brocklehurst et al.[73] have confirmed that the purified enzyme has quite similar properties. The most troublesome result of these studies has been the fact that the calcium dependence of protein kinase C seems to be substantially greater than the calcium dependence of release from the permeabilized cell systems. However, it remains possible that protein kinase C *in situ* may be under cooperative controls missing in the more purified states. Pocotte and Holz[14] have additionally suggested that phorbol ester could evoke phosphorylation of tyrosine hydroxylase, possibly via protein kinase C. The evidence for this ''housekeeping chore'' as an aspect of protein kinase C action, however, also remains indirect.

C. Metalloendoprotease Inhibitors and Secretion

Based on certain analogies to viral fusion, Strittmatter and co-workers[15] have recently proposed that metalloendoproteases might be generally involved in fusion of biological membranes. Evidence for this hypothesis is based on experiments in which inhibition of fusion in rat myoblast cultures was observed in the presence of certain oligopeptides, believed to be specific substrates and/or inhibitors of metalloendoproteases.

Indeed, the same oligopeptide inhibitors that inhibited fusion of myoblast cells,[15] neurotransmitter release at the mouse neuromuscular junction,[16] and histamine release from mast cells[17] were also found to interfere with catecholamine secretion from bovine adrenal chromaffin cells.[18] In mast cells and in chromaffin cells inhibition of secretion was dependent on both the type of stimulus employed and on the concentration of the respective agonists. Most effective inhibition of catecholamine release was observed when stimulating the chromaffin cells via the nicotinic receptor, while secretion evoked by other secretagogues, e.g., elevated extracellular K^+, Ba^{2+}, or veratridine, was much less susceptible to inhibition by the antagonist to metalloendoprotease activity.

This preferential inhibition of receptor-mediated stimulation by the oligopeptides casts doubt about their site of action, since it raises the possibility that these hydrophobic compounds might also exert nonspecific anticholinergic effects. Therefore, conclusions as to the involvement of metalloprotease activity in (exocytotic) membrane fusion, which are based on such inhibition studies, should be taken with great caution.

In the search for alternate cellular effects, we found that metalloprotease inhibitors which interfered with catecholamine secretion also modulated transmembranal calcium fluxes and the concentration of free intracellular Ca^{2+}.[74] Efflux of $^{45}Ca^{2+}$ was accelerated in the presence of the inhibitors, while at the same time the cytoplasmic-free Ca^{2+} was raised to a level, which by itself was not sufficient to cause catecholamine release. This elevated Ca^{2+} level, however, prevented a further increase in the cytoplasmic-free Ca^{2+} concentration upon addition of the various secretagogues.

Thus, with the above-mentioned cautionary note about possible nonspecific effects of the metalloprotease inhibitors in mind, we might speculate that metalloendoprotease activity is involved in regulating intracellular calcium homeostasis and/or calcium-dependent receptor inactivation. Rather than modifying a putative fusion protein, a metalloprotease activity might be required for the proteolytic cleavage of a protein, which itself might be involved in regulating the function of receptor-associated calcium channels.

III. CYTOSKELETAL PROTEINS

A. Interaction of Granules with Actin

The cytoplasm of chromaffin cells is composed of cytoskeletal elements including microfilaments, microtubules, and intermediate filaments,[19-21] and direct interactions between chromaffin granules and cytoskeletal elements have been observed by stereoelectron microscopy of cells embedded in water-soluble media.[22] We presumed there must be quite specific mechanisms regulating granule-cytoskeletal interaction, since the granules clearly move from one region of the cell to another.

Actin and myosin have been widely described in chromaffin cells,[23-30] and actin may actually be associated with the granule membrane. However, in highly purified chromaffin granule membranes, Zinder et al.[31] showed that the band comigrating with authentic actin was not actin. They used a highly specific fingerprint analysis of peptides eluted from the protein in the band. By contrast, a band comigrating with actin on gels of purified plasma membranes from chromaffin cells was, by the same criterion, indeed, actin. The basis of this difference of opinion about actin being on granules may rest with the fact that granule membranes analyzed by the different groups may have differed in purity. However, this does not mean that no association occurs in vivo and, as will be evident, such studies with less pure granule membrane preparations may prove to be more relevant to the in vivo situation.

Indeed, actin interacts with granule membranes when mixed together, as originally described by Burridge and Phillips.[24] Wilkins and Lin[32] also reported that they could detect stable oligomers of actin on granule membranes, using binding of radiolabeled cytochalasin B as an assay. Wilkins and Lin also suggested that these oligomers might be nuclei for the subsequent assembly of actin filaments.

However, Fowler and Pollard[33,34] found that F-actin could interact with highly purified granule membranes depleted of endogenous actin. The technique they used was low shear, falling-ball viscometry, in which the chromaffin granule membranes cross-linked F-actin and thus raised the viscosity of the solution. This interaction was inhibited by calcium, with 50% inhibition occurring at 0.2 μM free Ca^{2+}. Anticalmodulin and antisynexin drugs such as trifluoperazine and promethazine had no influence on this activity. On the other hand, trypsin treatment of membranes blocked the cross-linking activity, indicating that the actin binding site might be protein in nature.

α-Actinin has historically been detected in preparations of granule membranes.[35,36] α-Actinin is a component of Z bands of muscle, implicated by some investigators in the interaction of actin with the organelle and was thus considered a reasonable candidate for the F-actin binding site on granule membranes. However, the purified membranes used by Fowler and Pollard[33,34] were prepared under magnesium-free, low ionic strength conditions designed to elute out α-actinin.[27] Aunis and Perrin[37] have verified that α-actinin is indeed removed and have proposed instead that a spectrin-like protein (fodrin) might be the true granule membrane component responsible for binding F-actin (see Chapter 7 for more details).

The sensitivity to calcium may be the most important property of the actin binding site on granule membranes in terms of exocytosis regulation. It is possible that under resting conditions of cell calcium (≤ 0.1 μM free calcium) granules might be relatively immobilized, attached to F-actin in the cytoplasm. However, upon elevation of cell calcium after stimulation, the interaction between F-actin and granules would break. Granules would then become free to interact with the cell membrane and undergo subsequent fusion processes.[38] Indeed, direct visual evidence for this notion may have been provided in recent experiments by Aunis and co-workers,[39] who found that fodrin was specifically localized around the plasma membrane, and that upon stimulation of cells the distribution of cytoimmunospecific fodrin appeared modified near the plasma membrane.

B. Influence of Cytoskeletal Drugs and Proteins

However, in spite of interesting chemical data linking cytoskeletal proteins to secretory vesicles and plasma membranes, evidence for relevance to actual secretion processes has been difficult to obtain. Early experiments, using anticytoskeletal drugs such as colchicine, vinblastine, or cytocholasin B, were compromised by the findings that inhibition of catecholamine release by these drugs may be predominantly due to their strong anticholinergic effects, rather than to specific action on microfilaments or microtubules.[40,41] Furthermore, the data gathered from the use of anticytoskeletal drugs in electrically permeabilized ''leaky'' cells have seemed to clearly contradict the hypothesis for a role of cytoskeletal proteins in Ca^{2+}-induced exocytosis.[4] Indeed, these drugs had little, if any, effect on release from permeabilized cells. However, the permeabilized cell as a model for secretion has a major deficiency recognized by most who have studied the system. By virtue of the permeabilization step the natural pathways of signal transduction across the plasma membrane are necessarily short circuited. This means that one cannot tell whether the processes now observed are truly of importance in the intact cell.

Recently, liposome-cell fusion has been employed to overcome the membrane permeability barrier in intact chromaffin cells.[42] Using this technique (see Chapter 9), microfilament-specific macromolecules, such as heavy meromyosin and DNAse I, have been introduced efficiently into viable isolated chromaffin cells. The outcome of these experiments indicate a modulatory role of the cytoskeleton exactly at the level of transmembranous signaling, a site hitherto inaccessible using permeabilized cells. Following liposome-mediated injection of DNAse I, which is known to shift the G-F equilibrium of actin, Friedman and co-workers[43] observed an increase in catecholamine release concomitant with a significant sustained depolarization of the plasma membrane.

More recent experiments have shown similar effects for heavy meromyosin and its S_1 subfragment, but not if the same compounds were rendered mechanochemically inactive by poisoning them with N-ethyl maleimide (NEM).[42] DNAse I and heavy meromyosin, but not NEM-poisoned heavy meromyosin, also induced sustained membrane depolarization concomitant with a significant increase in the influx of sodium and calcium. In addition, Ca^{2+} (20 μM)-induced catecholamine release from digitonin permeabilized cells) was augmented in the presence of DNAse I and heavy meromyosin, but not upon addition of NEM heavy meromyosin. In contrast, addition of F-actin resulted in a marked decrease of Ca^{2+}-stimulated release from digitonin-permeabilized cells.[75]

These data thus indicate that these macromolecular reagents introduced by liposome fusion methods may have actions on membrane sites in addition to cytoskeletal sites, either directly or indirectly. Nonetheless, it is highly likely that specific location and dislocation of granules in the cytoplasm of chromaffin cells, after introduction of calcium, will prove to be a sensitive function of the easily detected interaction between granule membranes and F-actin. Indeed, Morita and Pollard[44] have recently shown that actin also stabilizes a granule ATPase activity against thermal denaturation, thus indicating more extensive effects of actin than mere binding. Actin and myosin may also prove to have concomitant actions on separate plasma membrane functions related to secretion.

IV. SYNEXIN

A. Synexin and Granule Aggregation

For exocytosis to occur the granule membrane must contact and fuse with the plasma membrane. Therefore, to understand the process one must explain how calcium can induce these essentially mechanochemical events. While calcium can act directly in a number of model systems, many biochemical processes depend on a calcium binding protein for specificity and sensitivity. In the case of exocytosis, we have proposed that synexin, or a synexin-like protein, might be a likely mediator of membrane contact and membrane fusion. Indeed, in the presence of calcium only this protein and its functional relatives promote these events when added to isolated chromaffin granules.

Synexin from adrenal medulla is a 47,000-dalton calcium binding protein that causes isolated chromaffin granules to aggregate to one another by pentalaminar membrane contact.[45,46] The activity can be measured by simply following the turbidity of a granule suspension at A 540 nm. Since secretion from chromaffin cells proceeds by both simple and compound exocytosis the formation of such contacts may indeed be of physiological relevance. In simple exocytosis the granule fuses with the plasma membrane, but in compound exocytosis chromaffin granules make contact and fuse with granule membrane remnants of previous exocytotic events remaining on the plasma membrane. This process may be of great advantage in chromaffin and other cells, since upon stimulation only limited movement of granules is necessary, and the resulting tunnels of fused granules may provide a pathway for controlled penetration of calcium-rich extracellular medium into the depths of the cell.

The exact mechanism of synexin-dependent granule aggregation is a source of some controversy. Creutz et al.[46] found that the calcium titration curve for granule aggregation coincided with that for calcium-dependent polymerization of synexin to form 50 × 100-Å rods. They suggested that calcium acted on synexin to form active polymers, which themselves caused granules to aggregate. Only calcium promoted both processes, and calcium-dependent polymerization of synexin has indeed also been used as part of a synexin purification scheme by Morris et al.[47] However, Morris and co-workers preferred the interpretation that calcium acted only on the granule membranes and that synexin promoted this calcium effect on a membrane site.

The site to which synexin binds on the granule membrane is similarly a source of controversy. Dabrow et al.[48] found that treatment of granules with proteolytic enzymes could block synexin action. They concluded that the synexin receptor was a protein. However, Hong et al.[49] reported that synexin caused aggregation and fusion of phosphatidyl serine vesicles in a calcium-dependent manner. The latter authors questioned the existence of a protein receptor and suggested instead that the receptor could be a lipid. In circumstantial support of this conclusion were findings, summarized by Creutz et al.,[50] that synexin was among a series of proteins, termed "chromobindins",[51] that had affinity for the lipid fraction of chromaffin granules when these lipids were bound to a sepharose affinity column.

Pursuing the concept that the synexin receptor might be a lipid, Hong et al.[52] investigated the ability of synexin to aggregate and fuse liposomes of defined composition. The specificity of synexin was manifest by the observation that calmodulin slightly inhibited phospholipid vesicle fusion induced by calcium, while other calcium binding proteins such as bovine prothrombin and its proteolipid fragment 1 had a strong inhibitory action. Synexin alone was able to lower the threshold for calcium-induced fusion of phosphatidic acid (PA): phosphatidylethanolamine (PE)::(1:3) liposomes from 1 mM Ca^{2+} to 100 μM Ca^{2+} and subsequently to approximately 10 μM Ca^{2+} in the additional presence of 1 mM Mg^{2+}. Hong et al.[52] believed that the mechanism of the fusion process depended upon formation of anhydrous Ca^{2+} complexes with acidic phospholipid head groups, and that synexin somehow promoted this effect.

However, one can question whether these liposome studies accurately model the actual interaction of synexin and chromaffin granules. For example, the specific dependence of this liposome interaction on calcium and its potentiation by magnesium are somewhat different from the observed action of synexin on intact chromaffin granules. Calcium and synexin only induce aggregation of granules, not aggregation and fusion as the combination does with PA/PE::(1:3) liposomes. Furthermore, synexin action does not need magnesium to achieve its threshold for granule aggregation at approximately 6 μM Ca^{2+}.[53] Finally, added magnesium neither potentiates nor inhibits calcium-activated synexin activity. Indeed, the critical variable affecting binding of synexin to chromaffin granule membranes, aside from calcium, is pH. The $K_{1/2}$ for calcium dependence of synexin binding to granule membranes is approximately 5 μM at neutral pH and rises as the pH declines.[53]

However, one element in the study by Hong et al.[52] was more reminiscent of the effects of synexin on native chromaffin granules. In a survey of a variety of phospholipids, the authors discovered that phosphatidylinositol (PI) profoundly inhibited fusion, while still allowing liposome aggregation, when it replaced phosphatidic acid in the PA/PE (1:3) liposomes. This was seen as a unique property of PI, possibly due to the inositol head group blocking access of synexin to the phosphate group of the phospholipid. However, when viewed from the perspective of the synexin reaction with granules, it is an intriguing possibility that phosphatidyl inositol may be the bona fide lipid receptor for synexin, rather than the other phospholipids.

B. Synexin and Granule Fusion

While calcium and synexin can only aggregate chromaffin granules, it is indeed possible to cause the granule aggregates to fuse. This is achieved by the addition of a small amount (5 μM) of arachidonic acid. Other fatty acids will also work so long as the fatty acid has a *cis* unsaturated bond(s).[54,55] Fusion can be followed by phase microscopy or by a reduction

Table 1
KNOWN SYNEXIN-LIKE OR RELATED PROTEINS

Protein	Source	Mol wt	Action	Ref.
Synexin	L, A, B, S	47 K	Agg. granules $K_{1/2}$ = 200 μM, Thr = 5 μM	45
ν-Synexin	↑ L, A	38 K	Agg. granules, at $[Ca^{2+}] \leq 1$ μM	1
Synhibin	L, ↑ A	67 K	Inhibits synexin, ν-synexin	65
Synexin II	A, L	56 K	Agg. granules in 800 μM Ca^{2+}	66
μ-Synexin	L, A, ↑ M	51 K	Similar to 47 K synx, only form in muscle	68
Calelectrin	Torp.	34 K	Agg. granules, $K_{1/2}$ = 200 μM	69
	mam B, A	32.5 K	Same	70
	mam B, A	67 K	Same	70

Note: L = liver; A = adrenal medulla; B = brain; S = spleen; M = muscle; torp = *Torpedo marmorata*; mam = mammalian; thr = threshold; ↑ = increased.

in the turbidity at A 540 nm. The relevance of this reaction to exocytosis in vivo rests in the fact that the fusion structures formed by treatment of granule aggregates with arachidonic acid are very similar to structures observed in secreting cells.[38] The fatty acid effect is also likely to be physiological, since arachidonic acid is rapidly produced by chromaffin cells stimulated to secrete catecholamines.[56,57] An important conclusion from these observations is that a phospholipase A_2-type activity may be important for secretion. Phospholipase A inhibitors do block secretion from chromaffin cells, but as Frye and Holz[58] have shown, phospholipase inhibitors also block calcium uptake, thereby rendering the effect on secretion presently uninterpretable.

C. Synexin Blockade by Phenothiazine Drugs

The synexin-dependent granule aggregation reaction is also quite sensitive to phenothiazine drugs, as are proteins such as calmodulin or processes such as local anesthesia. Trifluoperazine (TFP) blocks synexin-dependent granule aggregation at a concentration of approximately 4 μM when the reaction is carried out in a high (400 μM) calcium concentration.[59,60] However, promethazine (PMTHZ), a phenothiazine with low affinity (300 to 500 μM) for calmodulin and similarly, poor local anesthetic potency, was nearly as potent as TFP as a synexin blocker. In secreting chromaffin cells both TFP and PMTHZ were similarly potent inhibitors with ID_{50} values of approximately 1 μM,[60] doses well below those observed for TFP to block Ca^{2+} transport in some studies,[61,62] but not in others.[63] TFP and PMTHZ also inhibited glucose-induced insulin release from islets of Langerhans at similarly low doses, perhaps indicating some generality of the effect.[64] By presuming a parallel between granule aggregation and compound exocytosis we conclude that these data may provide some direct but obviously fragile support for the involvement of a synexin-like process in secretion.

D. Synexin-Like or Synexin-Related Proteins

A number of synexin-like or synexin-related proteins has been reported since the initial discovery of synexin in 1978. As shown in Table 1, there are at least eight in total. The etymology of synexin, from the Greek word "synexos" for a "meeting" or "coming together", has indicated proteins which, in the presence of calcium, cause chromaffin granules to aggregate. However, the next protein to be discovered was synhibin, a 67,000-dalton protein that "inhibited" synexin.[66] The inhibition appeared to be competitive with synexin, but seemed unrelated on the basis of monoclonal and polyclonal antibodies to synexin. Synhibin was initially found in liver, but was later found to be much more abundant in the adrenal medulla.

A number of synexin variants has been reported. Synexin II is a 56,000-molecular weight protein that has been found to be a minor contaminant of synexin preparations prepared from the adrenal medulla.[66] Like synexin, it aggregates chromaffin granules in the presence of calcium. The relation of synexin II to standard synexin is in question.

Most synexin preparations from adrenal medulla or liver also contains a minor 51,000-dalton component which is reactive with polyclonal antibodies to synexin, and which aggregates chromaffin granules with a strict calcium dependence similar to that for authentic synexin.[67] It is the only form of synexin found in striated muscle and has accordingly been named[68] μ-synexin. Its function in muscle is not known.

By contrast to secretion from chromaffin cells, other types of cells such as liver, parathyroid, and renin-secreting cells from the juxtaglomerular apparatus of the kidney secrete when cytosolic calcium concentration declines. Also, many fusion events certainly occur within the cytoplasm of chromaffin and other cell types where the ambient calcium concentration is low. We have recently reported the existence of a new synexin-like protein, called ν-synexin, which aggregates granules only in low calcium concentrations. Accordingly, it is assayed in 1 mM EGTA. In contrast to regular synexin, raising the calcium ion concentration blocks ν-synexin, while both strontium and barium are without effect. On the other hand, like authentic synexin granules aggregated by ν-synexin are fused by arachidonic acid. Furthermore, ν-synexin is blocked by both TFP and PMTHZ as well as by synhibin. The molecular weight is 38,000 daltons and is found in adrenal medulla, but to a greater extent in liver. Synexin and ν-synexin have some similarities in amino acid composition and have blocked (but unknown) N terminals. Polyclonal and monoclonal antisynexin antibodies do not react with ν-synexin. These details are summarized by Pollard et al.[1]

Yet a different group of synexin-like activities has been reported from Whittaker's laboratory, but given a different name, the "calelectrins". Initially, they were found in the electric organ of *Torpedo marmorata* and hence given this special name. The calelectrin from the electric organ was found by Walker[69] to be a protein of 35,000 mol wt, with aggregation properties similar to those for synexin. Later, Sudhof et al.[70] reported the discovery of calelectrins in mammalian tissues, including bovine liver and adrenal medulla, that were similar in properties to the *Torpedo* protein. However, in mammalian tissues the aggregating activities were associated with molecular species of both 32,500 and 67,000 mol wt. Antibodies to each calelectrin species reacted with each of the other two proteins, regardless of origin. In addition, like synexin, the calelectrins also caused phosphtidylserine liposomes to aggregate, indicating a possible common mechanism of action. However, according to Creutz,[76] calelectrins do not react with antibodies to authentic synexin. A caveat regarding the possible generality of these calelectrins must at this point be issued, since calcium binding proteins similar in molecular weight have been isolated from pig tissues, but found to be immunologically distinct.[71] However, no chromaffin granule aggregation studies on the porcine material have yet been reported.

V. CONCLUSIONS

In this chapter we have emphasized the possible roles of several cytoplasmic mediators of calcium action during the process of exocytosis. Actin and chromaffin granules form complexes in vitro which are calcium dependent. Very likely these complexes represent interactions which can occur in the cytosol of chromaffin cells. Synexin causes membrane contact and fusion events to occur, mirroring such events in secreting cells. Very likely synexin, or one of a burgeoning number of synexin-related proteins, represents the events involved in exocytosis. The compelling reason for considering these reactions relevant is that they involve calcium-regulated protein species within the cell which form structures in vitro that closely model structural events occurring in secreting chromaffin cells.

It has remained our concern that the calcium sensitivity of the synexin reaction might be "too high" ($K_{1/2}$ = 200 μM) to be relevant physiologically. However, there is a growing realization that compartmentalization of submembrane calcium activity during calcium influx can lead to local free calcium concentration calculated to be tens to hundreds of micromoles per liter in nerve terminals,[72] concentrations certainly relevant to activating authentic synexin or blocking v-synexin. Another reason for interest in synexin is the fact that in the chromaffin cell, at least, synexin or synexin-like proteins are considered so important as to warrant large amounts of a specific inhibitor, synhibin.

Other proteins, some quite popular (e.g., calmodulin; see Chapter 6), were simply mentioned in our introduction, but not pursued further. We ignored them in the text for the simple reason that presently they do nothing discernible to the model reactions we have studied for granule/cytoskeleton interaction or granule membrane contact and fusion. The future of these studies undoubtedly lies in developing the entire repertoire of interactions between granules and cytoskeletal elements, on the one hand, and synexin-related proteins on the other. It is our feeling that the length of Table 1 is only a hint of the possibilities to be realized over the next few years.

REFERENCES

1. **Pollard, H. B., Ornberg, R., Levine, M., Kelner, K., Morita, K., Levine, R., Forsberg, E., Brocklehurst, K. W., Duong, L., Lelkes, P. I., Heldman, E., and Youdim, M.,** Hormone secretion by exocytosis with emphasis on information from the chromaffin cell system, *Vitam. Horm.*, 42, 109, 1985.
2. **Beveridge, M. J.,** Inositoltrisphosphate and diacylglycerol as second messengers, *Biochem. J.*, 220, 345, 1984.
3. **Vergara, J., Tsien, R. Y., and Delay, M.,** Inositol-1,4,5 trisphosphate: a possible chemical link in excitation-contraction coupling in muscle, *Proc. Natl. Acad. Sci. U.S.A.*, 82, 6352, 1985.
4. **Fisher, S. K., Holz, R. N., and Agranoff, B. W.,** Muscarinic receptors in chromaffin cell cultures mediate enhanced phospholipid labeling but not catecholamine secretion, *J. Neurochem.*, 37, 491, 1981.
5. **Rojas, E., Pollard, H. B., and Heldman, E.,** Real time measurements of acetylcholine-induced release of ATP from bovine medullary chromaffin cells, *FEBS Lett.*, 185, 323, 1985.
6. **Forsberg, E. J., Rojas, E., and Pollard, H. B.,** Muscarinic receptor enhancement of nicotine-induced catecholamine secretion may be mediated by phosphoinositide metabolism in bovine adrenal chromaffin cells, *J. Biol. Chem.*, in press.
7. **Kao, L.-S. and Schneider, A. S.,** Muscarinic receptors in bovine chromaffin cells mediate a rise in cytosolic calcium that is independent of extracellular calcium, *J. Biol. Chem.*, 260, 2019, 1985.
8. **Oka, M., Isosaki, M., and Watanabe, J.,** Calcium flux and catecholamine release in isolated bovine adrenal medullary cells: effects of nicotinic and muscarinic stimulation, *Adv. Biosci.*, 36, 29, 1982.
9. **Nishizuka, Y.,** The role of protein kinase C in cell surface signal transduction and tumour promotion, *Nature (London)*, 308, 693, 1984.
10. **Knight, D. E. and Baker, P. F.,** The phorbol ester TPA increases the affinity of exocytosis for calcium in "leaky" adrenal medullary cells, *FEBS Lett.*, 160, 98, 1983.
11. **Morita, K., Brocklehurst, K. W., Tomares, S. M., and Pollard, H. B.,** The phorbol ester TPA enhances A23187 — but not carbachol — and high K^+-induced catecholamine secretion from cultured bovine adrenal chromaffin cells, *Biochem. Biophys. Res. Commun.*, 129, 511, 1985.
12. **Brocklehurst, K. W. and Pollard, H. B.,** Enhancement of Ca^{2+}-induced catecholamine release by the phorbol ester TPA in digitonin-permeabilized cultured bovine adrenal chromaffin cells, *FEBS Lett.*, 183, 107, 1985.
13. **Brocklehurst, K. W., Morita, K., and Pollard, H. B.,** Characterization of protein kinase C and its role in catecholamine secretion from bovine adrenal medullary cells, *Biochem. J.*, 228, 35, 1985.
14. **Pocotte, S. L. and Holz, R. W.,** Effects of phorbol ester on tyrosine hydroxylase phosphorylation and activation in cultured bovine adrenal chromaffin cells, *J. Biol. Chem.*, 261, 1873, 1986.
15. **Couch, C. and Strittmatter, W. J.,** Rat myoblast fusion requires metalloendoprotease activity, *Cell*, 32, 257, 1983.
16. **Baxter, D. A., Johnson, D., and Strittmatter, W. J.,** Protease inhibitors implicate metalloendoprotease in synaptic transmission at the mammalian neuromuscular junction, *Proc. Natl. Acad. Sci. U.S.A.*, 80, 4174, 1983.

17. **Mundy, D. I. and Strittmatter, W. J.**, Requirement for metalloendoprotease in exocytosis: evidence in mast cells and adrenal chromaffin cells, *Cell*, 40, 645, 1985.

18. **Lelkes, P. I., Brocklehurst, K. M., Morita, K., and Pollard, H. B.**, Metalloendoproteases in bovine adrenal chromaffin cells: their localization and involvement in stimulus-secretion coupling, in Proc. Catecholamine Symp., Colmar, France, 1984.

19. **Lazarides, E. and Weber, K.**, Actin antibody: the specific visualization of actin filaments in non-muscle cells, *Proc. Natl. Acad. Sci. U.S.A.*, 71, 2268, 1974.

20. **Fuller, G. M., Brinkley, B. R., and Boughter, J. M.**, Immunofluorescence of mitotic spindles using monospecific antibody against bovine brain tubulin, *Science*, 187, 948, 1975.

21. **Osborn, M., Franke, W. W., and Weber, K.**, Visualization of a system of filaments 7—10 nm thick in cultured cells of an epithelioid line (PtK 2) by immunofluorescence microscopy, *Proc. Natl. Acad. Sci. U.S.A.*, 74, 2490, 1977.

22. **Kondo, H., Wolosewick, J. J., and Pappas, G. D.**, The microtrabecular lattice of the adrenal medulla revealed by polyethylene glycol embedding and stereo electron microscopy, *J. Neurosci.*, 2, 57, 1982.

23. **Phillips, J. H. and Slater, A.**, Actin in the adrenal medulla, *FEBS Lett.*, 56, 327, 1975.

24. **Burridge, K. and Phillips, J. H.**, Association of actin and myosin with secretory granule membranes, *Nature (London)*, 254, 526, 1975.

25. **Trifaró, J. M. and Ulpian, C.**, Isolation and characterization of myosin from the adrenal medulla, *Neuroscience*, 1, 483, 1976.

26. **Creutz, C. E.**, Isolation, characterization and localization of bovine adrenal medullary myosin, *Cell Tissue Res.*, 178, 17, 1977.

27. **Aunis, D., Guerold, B., Bader, M. F., and Ciesielski-Treska, J.**, Immunocytochemical and biochemical demonstration of contractile proteins in chromaffin cells in culture, *Neuroscience*, 5, 2261, 1980a.

28. **Hesketh, J. E., Ciesielski-Treska, J., and Aunis, D.**, A phase-contrast and immunofluorescence study of adrenal medullary chromaffin cells in culture: neurite formation, actin and chromaffin granule distribution, *Cell Tissue Res.*, 218, 331, 1981.

29. **Lee, R. W. H., Mashyashi, W. E., and Trifaró, J. M.**, Two forms of cytoplasmic actin in adrenal chromaffin cells, *Neuroscience*, 4, 843, 1979.

30. **Lee, R. W. H. and Trifaró, J. M.**, Characterization of anti-actin antibodies and their use in immunocytochemical studies on the localization of actin in adrenal chromaffin cells in culture, *Neuroscience*, 6, 2087, 1981.

31. **Zinder, O., Hoffman, P. G., Bonner, W. M., and Pollard, H. B.**, Comparison of chemical properties of purified plasma membranes and secretory vesicle membranes from the bovine adrenal medulla, *Cell Tissue Res.*, 188, 155, 1978.

32. **Wilkins, J. A. and Lin, S.**, Association of actin with chromaffin granule membranes and the effect of cytochalasin B on the polarity of actin filament elongation, *Biochim. Biophys. Acta*, 642, 55, 1981.

33. **Fowler, V. M. and Pollard, H. B.**, Chromaffin granule membrane-F-actin interaction are calcium sensitive, *Nature (London)*, 295, 336, 1982.

34. **Fowler, V. M. and Pollard, H. B.**, In vitro reconstitution of chromaffin granule-cytoskeleton interactions: ionic factors influencing the association of F-actin with purified chromaffin granule membranes, *J. Cell Biochem.*, 18, 295, 1982.

35. **Jockusch, B. M., Burger, M. M., daPrada, M., Richards, J. G., Chaponnier, C., and Gabbiani, G.**, α-Actinin attached to membranes of secretory vesicles, *Nature (London)*, 270, 628, 1977.

36. **Bader, M. F. and Aunis, D.**, The 97-kd α-actinin-like protein in chromaffin granule membranes from adrenal medulla: evidence for localization on the cytoplasmic surface and for binding to actin filaments, *Neuroscience*, 8, 165, 1983.

37. **Aunis, D. and Perrin, D.**, Chromaffin granule membranes-F-actin interaction and spectrin-like protein of subcellular granules: a possible relationship, *J. Neurochem.*, 42, 1558, 1984.

38. **Pollard, H. B., Creutz, C. E., Fowler, V. M., Scott, J. H., and Pazoles, C. J.**, Calcium-dependent regulation of chromaffin granule movement, membrane contact, and fusion during exocytosis, *Cold Spring Harbor Symp. on Quant. Biol.*, 46, 819, 1982.

39. **Perrin, D. and Aunis, D.**, Stimulation induces alpha-fodrin reorganization in secretory cells, *Nature (London)*, 315, 589, 1985.

40. **Douglas, W. W. and Sorimachi, M.**, Colchicine inhibits adrenal medullary secretion evoked by acetylcholine without affecting that evoked by potassium, *Br. J. Pharmacol.*, 45, 129, 1972.

41. **Schneider, A. S., Cline, H. T., Rosenheck, K., and Sonnenberg, M.**, Stimulus-secretion coupling in isolated adrenal chromaffin cells: calcium channel activation and possible role of cytoskeletal elements, *J. Neurochem.*, 37, 567, 1981.

42. **Lelkes, P. I. and Friedman, J. E.**, Interaction of French-pressed liposomes with isolated bovine adrenal chromaffin cells. Characterization of the cell-liposome interactions, *J. Biol. Chem.*, 260, 1796, 1985.

43. **Harish, O. E., Levy, R., Rosenheck, K., and Oplatka, A.,** Possible involvement of actin and myosin in Ca^{2+} transport through the plasma membrane of chromaffin cells, *Biochem. Biophys. Res. Commun.*, 119, 652, 1984.

44. **Morita, K. and Pollard, H. P.,** Chromaffin granule-cytoskeleton interaction. Stabilization by F-actin of ATPase in purified chromaffin granule membranes, *FEBS Lett.*, 181, 195, 1985.

45. **Creutz, C. E., Paxoles, C. J., and Pollard, H. B.,** Identification and purification of an adrenal medullary protein (Synexin) that causes calcium-dependent aggregation of isolated chromaffin granules, *J. Biol. Chem.*, 253, 2858, 1978.

46. **Creutz, C. E., Pazoles, C. J., and Pollard, H. B.,** Self-association of synexin in the presence of calcium: correlation with synexin-induced membrane fusion and examination of the structure of synexin aggregates, *J. Biol. Chem.*, 254, 553, 1979.

47. **Morris, S. J., Hughes, J. M. X., and Whittaker, V. P.,** Purification and mode of action of synexin: a protein enhancing calcium-induced membrane aggregation, *J. Neurochem.*, 39, 529, 1982.

48. **Dabrow, M., Zaremba, S., and Hogue-Angeletti, R. A.,** Specificity of synexin-induced chromaffin granule aggregation, *Biochem. Biophys. Res. Commun.*, 96, 1164, 1980.

49. **Hong, K., Duzgunes, N., and Papahadjopoulos, D.,** Role of synexin in membrane fusion. Enhancement of calcium-dependent fusion of phospholiipid vesicles, *J. Biol. Chem.*, 256, 3641, 1981.

50. **Creutz, C. E., Dowling, L. G., Sando, J. J., Vilbar-Palasi, C., Whipple, J. H., and Zaks, W. J.,** Characterization of the chromobindins: soluble proteins that bind to the chromaffin granule membrane in the presence of Ca^{2+}, *J. Biol. Chem.*, 258, 14664, 1983.

51. **Creutz, C. E.,** Secretory vesicle-cytosol interactions in exocytosis: isolation by Ca^{2+}-dependent affinity chromatography of proteins that bind to the chromaffin granule membrane, *Biochem. Biophys. Res. Commun.*, 103, 1395, 1981.

52. **Hong, K., Duzgunes, N., and Papahadjopoulos, D.,** Modulation of membrane fusion by calcium binding proteins, *Biophys. J.*, 37, 297, 1982.

53. **Creutz, C. E. and Sterner, D. C.,** Calcium dependence of the binding of synexin to isolated chromaffin granules, *Biochem. Biophys. Res. Commun.*, 114, 355, 1983.

54. **Creutz, C. E.,** Cis-unsaturated fatty acids induce the fusion of chromaffin granules aggregated by synexin, *J. Cell Biol.*, 91, 247, 1981.

55. **Creutz, C. E. and Pollard, H. B.,** A cell-free model for protein-lipid interactions in exocytosis, *Biophys. J.*, 37, 119, 1981.

56. **Hotchkiss, A., Pollard, H. B., Scott, J., and Axelrod, J.,** Release of arachidonic acid from adrenal chromaffin cell cultures during secretion of epinephrine, *Fed. Proc.*, 40, 256, 1981.

57. **Frye, R. A. and Holz, R. W.,** The relationship between arachidonic acid release and catecholamine secretion from cultured bovine adrenal chromaffin cells, *J. Neurochem.*, 43, 146, 1984.

58. **Frye, R. A. and Holz, R. W.,** Phospholipase A_2 inhibitors block catecholamine secretion and calcium uptake in cultured bovine adrenal medullary cells, *Mol. Pharmacol.*, 23, 547, 1983.

59. **Creutz, C. E., Scott, J. E., Pazoles, C. J., and Pollard, H. B.,** Further characterization of the aggregation and fusion of chromaffin granules by synexin as a model for compound exocytosis, *J. Cell Biochem.*, 18, 87, 1982.

60. **Pollard, H. B., Scott, J. H., and Creutz, C. E.,** Inhibition of synexin activity and exocytosis from chromaffin cells by phenothiazine drugs, *Biochem. Biophys. Res. Commun.*, 113, 908, 1983.

61. **Koenigsberg, R. L., Cote, A., and Trifaró, J. M.,** *Neuroscience*, 7, 2277, 1982.

62. **Baker, P. F. and Knight, D. E.,** Calcium control of exocytosis and endocytosis is bovine adrenal medullary cells, *Philos. Trans. R. Soc. London Ser. B*, 296, 83, 1981.

63. **Wada, A., Yanagihara, N., Izumi, F., Salcurai, S., and Kobayashi, H.,** Trifluoperazine inhibits ^{45}Ca uptake and catecholamine secretion and synthesis in adrenal medullary cells, *J. Neurochem.*, 40, 481, 1983.

64. **Sussman, K., Pollard, H. B., Leitner, J. W., Nesher, R., Adler, J., and Cerasi, E.,** Differential control of insulin secretion and somatostatin receptor recruitment in isolated pancreatic islets, *Biochem. J.*, 214, 225, 1983.

65. **Pollard, H. B. and Scott, J. H.,** Synhibin: a new calcium dependent membrane-binding protein that inhibits synexin-induced chromaffin granule aggregation and fusion, *FEBS Lett.*, 150, 201, 1982.

66. **Odenwald, W. F. and Morris, S. J.,** Identification of a second synexin-like adrenal medullary and liver protein that enhances calcium-induced membrane aggregation, *Biochem. Biophys. Res. Commun.*, 112, 147, 1983.

67. **Scott, J. H., Kelner, L. K., and Pollard, H. B.,** Purification of synexin by pH step elution from chromatofocusing media in the absence of ampholines, *Anal. Biochem.*, 149, 163, 1985.

68. **Dowling, L. G. and Creutz, C. E.,** Comparison of synexin isotypes in secretory and non-secretory tissues, *Biochem. Biophys. Res. Commun.*, 132, 382, 1985.

69. **Walker, J. H.,** Isolation from cholinergic synapses of a protein that binds to membranes in a calcium dependent manner, *J. Neurochem.*, 39, 815, 1982.

70. **Sudhof, T. C., Ebbecke, M., Walker, J. H., Fritsche, U., and Bonstead, C.,** Isolation of mammalian calelectrins: a new class of ubiquitous Ca^{2+}-regulation proteins, *Biochemistry*, 23, 1103, 1984.
71. **Shadle, P. J., Gerke, V., and Weber, K.,** Three Ca^{2+}-binding proteins from porcine liver and intestine differ immunologically and physicochemically and are distinct in Ca^{2+} affinity, *J. Biol. Chem.*, 260, 16354, 1985.
72. **Simon, S. M. and Llinas, R. R.,** Compartmentalization of the submembrane calcium activity during calcium influx and its significance in transmitter release, *Biophys. J.*, 48, 485, 1985.
73. **Brocklehurst, K. W., Lee, G., and Pollard, H. B.,** Rapid purification and properties of protein kinase C from bovine adrenal medulla, *Biosci. Rep.*, in press.
74. **Lelkes, P. I. and Pollard, H. B.,** unpublished observations.
75. **Friedman, B. and Lelkes, P. I.,** unpublished observations.
76. **Creutz, C.,** personal communication, 1985.

Chapter 9

LIPOSOME-MEDIATED INTRODUCTION OF MACROMOLECULES INTO ISOLATED BOVINE ADRENAL CHROMAFFIN CELLS

A Critical Evaluation of the Technique and its Biological Implications

Peter I. Lelkes

TABLE OF CONTENTS

I. INTRODUCTION

Intracellular signal transduction constitutes one of the most complex stages in stimulus-secretion coupling. A variety of different biochemical pathways have been identified over the past few years, and yet we are still far from assembling these mosaics into a more comprehensive picture.[1]

Our current difficulties to fully account for the cytosolic determinants in the stimulus secretion cascade derive, at least, in part, from the fact that the cytoplasm of *intact* cells is, per definition, not directly accessible to experimental manipulation. Ingenious techniques have been established in recent years to overcome the membrane permeability barrier, to ascertain direct access to the cytoplasm, and to create a model system for exocytosis.[2-5] By now a wealth of mostly consistent, but also partially contradicting data has emerged from studies of catecholamine release in permeabilized chromaffin cells, which seem to define the minimal cytosolic requirements for secretion.[6]

Permeabilized cells might be a suitable model for studying exocytosis. However, a major drawback in using these systems to identify cytoplasmic parameters which may govern stimulus-secretion coupling is that by virtue of the permeabilization process, all "natural" pathways of stimulus transduction across the plasma membranes are short circuited.[2] In order to study intracellular signaling under *in situ* conditions, the cells have to retain their structural and functional integrity following manipulation of the membrane permeability barrier. The techniques available today to achieve this goal are either the physical microinjection[7] and/or patch clamping[8] of single cells, or the fusion of entire cell populations with microscopic carriers, such as Sendai virus,[9] erythrocyte ghosts,[10] or liposomes.[11] The former techniques are extremely demanding and require a considerable technical setup. In addition, since they involve single cells, it is quite difficult to correlate, for example, electrophysiological and biochemical data from the same preparation.[8] In contrast, manipulations of the cytoplasm by fusing entire cell populations with appropriate vectors permit concomitant analysis of the effects, e.g., by analyzing microscopic, biochemical, as well as electrophysiological data.[11]

Recently, Trifaró and Kenigsberg[12,13] described in detail the techniques to fuse erythrocyte ghosts with bovine adrenal chromaffin cells in culture. The physiological effects of anti-calmodulin antibodies, microinjected by this technique, and the implications for the role of calmodulin in stimulus secretion coupling are discussed in Chapter 6.

This chapter has a dual purpose. First, I want to critically review the interaction of liposomes with isolated mammalian cells, in particular, with chromaffin cells: how can these interactions be assayed and quantified, which type of interactions prevail, and how feasible is such a liposomal delivery system? In the second part of this chapter I shall discuss two examples for such liposome-mediated modifications of the cytoplasm of chromaffin cells in suspension and describe how the exocytotic secretion of catecholamines can be modulated by (1) manipulating the cytoplasmic ionic composition and (2) by altering the state of assembly of cytoskeletal proteins such as the microfilaments.

II. ANALYSIS OF THE TECHNIQUE

A. How to Choose the Right Liposomal Vector

The main goal for using liposomal vectors is to transport biologically relevant (macro)molecules across the membrane permeability barrier of *intact* cells, preferentially by membrane fusion.[14] The right choice of a suitable liposomal carrier is essential, inasmuch as it has to ensure (1) that those macromolecules retain their biological activity during the encapsulation procedure, (2) that the entrapment efficiency into the liposomes is reasonable, and (3) that the interaction of the carrier vesicle with the target cell will be efficient enough to provide for sizeable delivery of the liposomal contents into the cytoplasm.

The first two issues, retention of biological activity and entrapment efficiency, are closely interrelated. Some of the more recent techniques to prepare liposomes will yield very high entrapment; however, the techniques employed (e.g., admixing of aqueous protein solutions and lipids contained in an organic solvent and/or sonication) will result in irreversible denaturation and/or deactivation of the proteins.[15] An acceptable compromise between vesicle size, entrapment efficiency, and gentle methodology is achieved by high pressure extrusion of a preswollen multilamellar liposome suspension in a French press.[16,17] In addition to essentially leaving the biological activities of many sensitive macromolecules intact, this technique produces extremely stable liposomes in terms of passive leakage of their aqueous contents and their susceptibility to membrane destabilization by exogenous compounds such as serum proteins.[18]

The efficacy of liposomal delivery depends primarily on the target cells chosen and on the composition of the carrier vesicle. Cells, in which the plasma membrane is in a quiescent state, e.g., erythrocytes, are essentially refractive to spontaneous fusion with liposomes.[19] Other systems, like secretory cells, which have an increased turnover of their plasma membranes, e.g., due to the perpetual cycles of exocytosis-endocytosis, appear to be more prone to fusion with liposomes, probably due to local and temporal instabilities of the cell surface. These instabilities might even be augmented, when working with freshly isolated cells or cells in suspension culture that have not fully regenerated their cell surface coat following the traumatic isolation with proteolytic enzymes. In contrast to their mode of interaction with freshly isolated chromaffin cells, liposomes will adsorb *reversibly* to numerous cells kept in tissue culture.[14,20] Internalization, if observed at all, will predominantly occur via receptor-mediated endocytosis.[21] Fusion between liposomes and mammalian cells in culture seems to be a rather rare event.[14,21]

In formulating a suitable liposomal carrier for fusion with intact cells in suspension, the lipid composition of the vesicles seems to play an important, but not as crucial a role, as might have been inferred from the studies of liposome-liposome interactions.[22] Negatively charged lipids are required to ensure divalent cation-induced fusion of lipid vesicles.[23] In contrast, a net negative charge on the liposome surface might enhance attachment of liposomes to planar lipid bilayers[24] or to biological membranes;[25] the fusion proper can also occur with neutral liposomes and in the absence of divalent cations, e.g., calcium.[25,26] The mechanism of the fusion reaction is not well understood and might comprise membrane deformation due to osmotic stresses and/or hydration forces.[24,26]

B. How to Assay Cell-Liposome Interactions

To visualize and qualitatively estimate the extent of the interactions, liposomes are loaded with fluorescent (macro)molecules and incubated with the cells for various periods of time. Aliquots of the washed and resuspended cells are then visualized microscopically and/or examined via a fluorescence-activated cell sorter, to determine the fraction of the cells that have incorporated liposomal contents.[11,27,28]

The extent of cell-liposome interactions is quantified and their mechanism(s) are characterized by a combination of radiotracer and fluorescent techniques.[11,29] Liposomes can be simultaneously labeled with ^{14}C-tagged phospholipids and ^{3}H- or ^{125}I-labeled water-soluble markers for their lipid bilayer membranes and their cytoplasmic contents, respectively. Since the specific activities are known, one can easily quantify the liposomes in association with the cell. However, the figures do not provide any mechanistic information, since they do not differentiate between liposomes that are merely attached to the cell surface and those that actually have transferred their contents into the cytoplasm. Liposomes that are loosely adsorbed can be removed by either extensive washing, addition of "cold", nonlabeled liposomes, or by centrifugation through an oil layer in a microfuge. These procedures are very effective in removing >99% of reversibly attached liposomes from a variety of cell surfaces, e.g., from erythrocytes,[19] but *not* from isolated chromaffin cells in suspension.[11]

To quantify the fraction of liposomes that actually have fused with the plasma membranes and injected their contents into the cytoplasm, fluorescent techniques have to be employed. These assays are based on concentration-dependent relief of self-quenching[11,21,28,29] and/or changes in the efficiency of fluorescence resonance energy transfer.[11,25,29,30] Some of these techniques can provide real-time measurements of the fusion reaction by either monitoring dilution of the aqueous liposome contents into the cytosol and/or mixing of the membranes. Quantitation of the mixing of aqueous contents is hampered by the intrinsic leakiness of the carrier liposomes which can be even more accentuated by the detergent-like action of the cell surface.[31]

Leakage is a general problem in using liposomal carriers both in vitro as well as in vivo, and much effort has been spent to minimize loss of the liposomal contents before they can be delivered to their target sites.[32] By carefully choosing the lipid composition and employing special liposome preparational techniques, destabilization of the liposomal membrane by plasma proteins or upon contact with cell surfaces can be significantly reduced.[18,33,34] In addition, appropriate marker molecules (e.g., fluorescence tagged macromolecules, or hydrophilic multicarboxylated fluorescein derivatives, such as 5(6)-carboxyfluorescein[28,29,35-37] or calcein[38]) are essential for truly assessing passive leakage of vesicle contents. In all these assays, the fluorescent probes are incorporated into the liposomes at self-quenching concentrations. Upon fusion of the liposomes with the cell membranes, the vesicle contents will be diluted into the cytoplasm, or leak out into the medium, to make life more complicated. In any case, an increased fluorescence signal is observed due to the relief of the fluorescence quenching.

Based on a similar rationale, liposome cell interactions can also be assayed following the mixing of membrane constituents.[30] A matching pair of headgroup-labeled fluorescent phospholipids[30,39-42] or lipid analogs[29] is incorporated at high enough concentrations (approximately 0.5 mol%) into the liposomes, to ensure effective Förster-type fluorescence resonance energy transfer (RET). Upon interaction with the cells, the fluorescence will remain unaltered, if the liposomes do not fuse with the plasma membrane; alternately, if fusion does occur, an increase in fluorescence is observed due to the relief of the self-quenching or to a decrease in the efficiency of energy transfer. In most situations, a mixed fusion/adherence is the prevalent mode of cell-liposome interactions; in this case the degree of fusion can be evaluated following the techniques and formalism outlined by Pagano and co-workers.[41]

Like the fluorescence assay monitoring mixing (dilution) of the vesicle contents, this technique also has its own pitfalls. The validity of the RET assay depends heavily on the assumption that the probes are *nonexchangeable*, i.e., they will not spontaneously diffuse out of the liposomal bilayer membranes and be inserted in the cells' plasma membranes. Experimental verification of this criterion is crucial and has to be performed for each new probe and/or new target membrane under consideration. Nonexchangeability of probes depends on their structure and, hence, on their physicochemical properties.[43] Lipid analogs with two fatty acid chains are more likely to be nonexchangeable than those containing a single acyl chain; the longer the fatty acid tails, the firmer the probes will rest in the bilayer membranes. Saturated fatty acids cause less perturbance of the bilayer matrix than unsaturated ones or fatty acid chains containing covalently coupled probes.[43] Therefore, headgroup-labeled phospholipids[30,41] or lipid analogs containing long saturated acyl chains, e.g., stearic acid,[29] seem to be suitable probes for studying membrane fusion. However, the rate of exchangeability for these probes is also dependent on the structure of the fluorescent molecule attached to the lipids and residing at the lipid/water interface. Thus, rhodamine-coupled phospholipids appear to be less exchangeable than the fluorescein-tagged ones.[44]

Recently, Hoekstra and collaborators[45] introduced a novel nonexchangeable probe, octadecylrhodamine, that in contrast to previously used fluorescent phospholipid analogs, can also be inserted into cell membranes. Based on concentration-dependent self-quenching of the probe, this technique can be applied as a complementary means to the RET assay for lipid vesicle fusion, or to monitor fusion between membrane vesicles of biological origin.

Another independent means to ascertain the mechanism of interaction between liposomes and chromaffin cells is to follow the fluorescence recovery after photobleaching (FRAP).[46] A nonexchangeable probe is inserted at low concentrations into the liposomal bilayer and, upon fusion with the cells, will become freely diffusible in the plasma membrane; its lateral diffusion coefficient is practically indistinguishable from the same probe inserted directly into the plasma membrane. In contrast, those probe molecules that are located in the adhered, but nonfused, liposomes sticking to the plasma membrane will be irreversibly bleached by a single pulse of the exciting laser beam. The fractional recovery of fluorescence bleaching following repeated laser pulses is a measure of the percent of liposomes that actually have fused.[11]

C. Characterization and Quantitation of Liposome-Chromaffin Cell Fusion[11,25]

For the majority of the experiments reviewed in this chapter, bovine adrenal chromaffin cells were prepared by collagenase digestion and kept in suspension culture, according to Schneider et al.[47] The standard liposomes used in most of our studies were small unilamellar vesicles composed of phosphatidylcholine (PC), phosphatidylserine (PS), and cholesterol (Chol), at a molar ration of 7:1:2 and prepared by high pressure extrusion in a French press.

The various procedures for analyzing the interactions between the cells and liposomes are outlined in the flowchart in Figure 1. The cells and the purified liposomes are coincubated according to the experimental protocol. For further studying cell physiological responses following interaction with the liposomes, the cells are repeatedly washed by low speed centrifugation to remove noninteracted liposomes. Comparison of the catecholamine contents of the first supernatant (viz., the incubation medium) to that of untreated controls will yield information whether or not the transfer of the liposomal contents will modulate any of cytoplasmic determinants related to the secretory process, e.g., the ionic composition of the cytosol, function of specific cytoplasmic proteins, etc. Following the incubation, the cells are analyzed in terms of, e.g., their viability, their response to nicotinic stimulation, and their membrane potential. This separation procedure allows further processing of the cells and is more gentle than the alternate separation technique. High speed centrifugation of the cells through an inert oil layer[48] will also effectively remove noninteracted liposomes, however, this technique is recommended only for determination of cell-associated radioactivity, when studying interactions with radiolabeled liposomes.

To visualize liposome-mediated transfer of macromolecules into the cytoplasm, the isolated chromaffin cells are incubated for 30 min with liposomes loaded with fluorescence-tagged macromolecules. Analysis of the washed cells by flow cytometry and fluorescence microscopy indicates that more than 95% of all cells in a given suspension appear fluorescent (Figure 2); the perinuclear exclusion, seen in the micrograph, and the increase in the cell-associated fluorescence with time measured by flow cytometry confirm the intracellular origin of the fluorescence signal.

When liposomes, radiolabeled for both their membrane lipids and for their aqueous contents, are incubated with chromaffin cells, the cell-associated radioactivity will yield information on the overall extent of the interaction only, without differentiating the possible mechanism of these interactions. These interactions were found to be concentration dependent and saturable. At quasiequilibrium, the apparent K_m was approximately $5 \times 10^{-7}\,M$ lipid per 10^6 cells.[11] This value is very similar to K_m values found for the interaction of the same type of liposomes with rat basophilic leukemia cells[49] or for the binding of phosphatidylserine liposomes to rat peritoneal mast cells.[50] The association of liposomes with isolated chromaffin cells in suspension appears to be essentially irreversible; once bound to the cell surface,

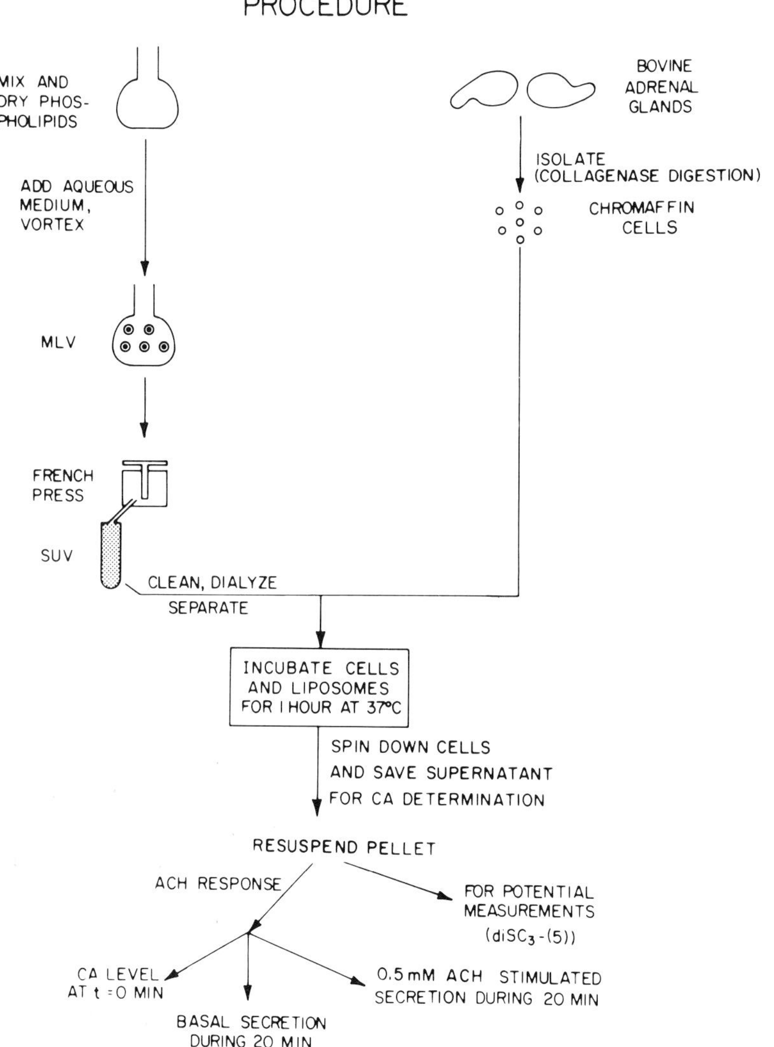

FIGURE 1. Flow chart outlining the experimental procedures employed for assaying the interactions between French pressed liposomes and isolated bovine adrenal medullary chromaffin cells and for determining cell physiological responses. For details, see text.

these vesicles are refractive to removal by pulse chasing or by trypsinization,[29] suggesting that the involvement of membrane proteins in this high affinity binding is only limited. Such an irreversible association is quite unique for chromaffin cells *in suspension,* since with *cultured* cells (including chromaffin cells) chasing will lead to significant loss of cell-associated liposomes.[51] In line with previous reports in other systems,[52] binding to chromaffin cells in suspension as well as in primary tissue culture is significantly reduced in the presence of serum albumin in the incubation mixture.[25,53]

The temperature dependence of the interaction of liposomes with mammalian cells was generally found to be pronounced.[14,54,55] For chromaffin cells, there is a break in the Arrhenius plot at ~18°C, reminiscent of the phase transition temperature of phosphatidylserine, which is present in both the liposomes and in the chromaffin cell plasma membranes. The radiotracer studies further showed that in contrast to several other systems,[54] the interactions of French pressed liposomes with chromaffin cells were not significantly suppressed by inhibitors of

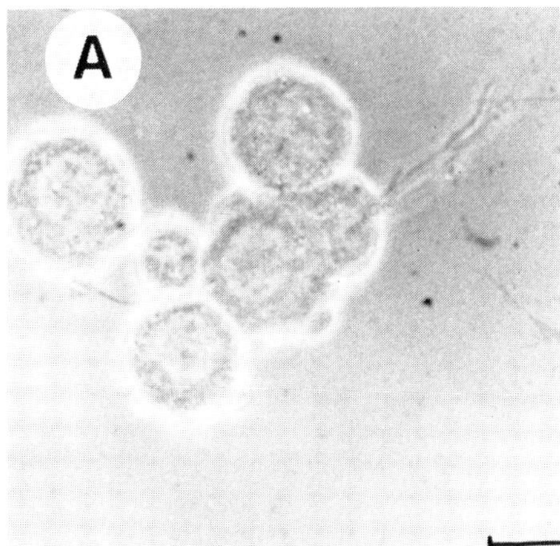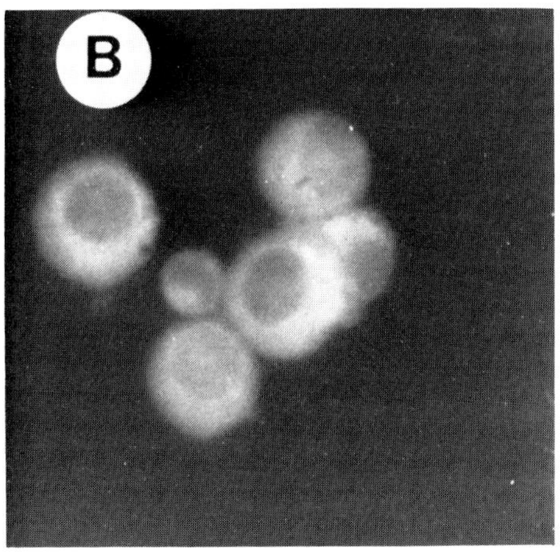

FIGURE 2. Liposome-mediated introduction of rhodamine-labeled bovine serum albumin into isolated bovine adrenal medullary cells in suspension. Cells were incubated for 30 min at 37°C, washed twice, and viewed at magnification × 800. The micrographs represent the same group of cells, taken in phase contrast (A) and in the fluorescence (B) mode. (From Lelkes, P. I. and Friedman, J. E., *J. Biol. Chem.*, 260, 1796, 1985. With permission.)

both the aerobic metabolism and those of glycolysis. Similar results previously obtained for other types of liposomes and different target cells were taken as indications for the occurrence of membrane fusion, which is supposedly governed by interbilayer contact phenomena, rather than by processes involving cellular metabolism.[14,54,55]

Radiotracer data encompass global information on all the liposomes that have interacted with the cells, irrespective of whether they are tightly adsorbed to the cell surface or whether they have actually fused with the plasma membranes. Therefore, other independent assays were needed to quantify simultaneously both the extent of the overall cell-liposome interactions and also to assess the actual number of liposomes fused with the plasma membranes. Fluorescence resonance energy transfer (RET) was used as a routine assay for the reasons discussed above; some of the results from these experiments were verified by studying the fluorescence recovery after photobleaching (FRAP).[11,46]

Using these fluorescence assays, chromaffin cell-liposome interactions were screened for a number of experimental parameters, such as incubation conditions, liposome compositions, etc.[25,56] Some of the results are summarized in Table 1. As already found in the radiotracer studies, the total amount of liposomes interacting with the cells increased with increasing liposome concentration. For effective cell-liposome adhesion to occur, the experiments had to be carried out in bovine serum albumin-free medium in the presence of calcium. An up to fivefold increase in the amount of cell-adsorbed liposomes was achieved by adding 100 μM lanthanum to the incubation medium; this trivalent cation, which is known to bind strongly to negatively charged liposomes,[57] was found to effectively promote cell-liposome interactions.

In contrast to the dose dependence of cell-liposome adhesion, the fraction of *fused* vesicles was found to be inversely proportional to the concentration of liposomes in the incubation medium. Raising the external liposome concentration by two orders of magnitude resulted in a parallel increase in the number of cell-associated liposomes; however, the fraction of liposomes that had actually fused with the plasma membranes was reduced from approximately 60% to approximately 20% (Figure 3). The mechanism of fusion seems to be independent of how cell adhesion was promoted, since for a given number of cell-associated liposomes the efficiency of membrane fusion remains constant irrespective of the incubation conditions.

Table 1
ANALYSIS OF CELL-LIPOSOME INTERACTIONS BY FLUORESCENCE RESONANCE ENERGY TRANSFER[25]

Liposome conc. (nmol lipid) & exp. conditions during incubation[a]	Amt. of cell-associated vesicles (% of total)	Fraction of liposomes fused (% of associated liposomes)
20, B1 + Ca^{2+}	5.0 ± 0.8	60 ± 2
1000, B1 + Ca^{2+}	0.4 ± 0.03	35 ± 3
20, B1 + Ca^{2+} + 0.5% BSA	2.7 ± 0.2	60 ± 5
Trypsin-treated cells 20, B1 + Ca^{2+}	5.5 ± 0.9	58 ± 2
20, B1 + 0.5 mM EGTA	2.7 ± 0.2	62 ± 2
20, B1 + 100 μM La^{3+}	4.6 ± 0.5	52 ± 2
20, B1 + Ca^{2+} + 100 μM La^{3+}	21.5 ± 0.8	40 ± 5

[a] French pressed liposomes, composed of phosphatidylcholine/phosphatidylserine/cholesterol at a molar ratio of 7:1:2, were incubated with 10^6 bovine adrenal medullary cells in physiological buffer 1 (B1)[47] containing 2.2 mM Ca^{2+}, as listed in the table. The extent of cell liposome interactions and the quantitation of membrane fusion were calculated according to Pagano et al.[41] Data are compiled from Dardik.[25] See text for details.

The major experimental parameter modulating chromaffin cell-liposome interactions appears to be the composition of the vesicles.[25,56] Similar conclusions have been drawn from the numerous studies on liposome-liposome fusion.[22] However, the liposome-cell membrane fusion may be governed by other (protein-mediated?) mechanisms as well, which do not fit the conventional picture of liposome fusion. While it is clear that calcium will certainly promote cell-liposome adhesion, the data seem to indicate that calcium is not required for the fusion step to occur. This observation is in line with recent findings on fusing negatively charged liposomes to biological membranes, e.g., chromaffin granule ghosts.[58] Furthermore, increasing the phosphatidylserine contents of liposomes results in enhanced calcium-induced vesicle-vesicle fusion.[59] With chromaffin cells as target membranes, however, PS-rich liposomes will show increased cell adherence only, while the fraction of fused vesicles is greatly reduced in comparison to liposomes containing 10 mol% PS only.[25,56] Also, in contrast to the findings for liposome-liposome fusion,[60] replacement of egg phosphatidylcholine by phosphatidylethanolamine does not result in significant changes in the number of cell-associated or fused liposomes, respectively, as compared to the standard liposomes, neither does the use of saturated phosphatidylcholines or a variation in the length of their fatty acid chains.[61] These latter parameters seem to have significant physiological implications for catecholamine release from liposome-treated cells.[18]

Optimal cell adherence and fusion were obtained for liposomes containing 30 mol% or more of cardiolipin. This phospholipid bears two net negative charges (at physiological pH); nevertheless, liposomes prepared from pure cardiolipin require higher calcium concentrations for fusion than those composed of pure phosphatidylserine, which bears one negative charge.[62] In mixed liposomes, however, cardiolipin appears to be superior to phosphatidylserine, in particular, in fusing with membranes of biological origin.[58,63] In comparison to the standard liposomes (PC/PS/Chol, 7:1:2) the efficacy of cell-liposome fusion was found to increase by up to fourfold, when using cardiolipin-containing vesicles in the presence of 2.2 mM calcium and 100 μM lanthanum.

The conclusion from the fluorescence experiments is that liposomes interact with freshly isolated chromaffin cells in suspension by a stepwise mechanism. Initially, the liposomes will strongly adhere to cell surface (preferentially to the lipids?). Subsequently, a fraction of those adsorbed vesicles is internalized, predominantly by membrane fusion rather than

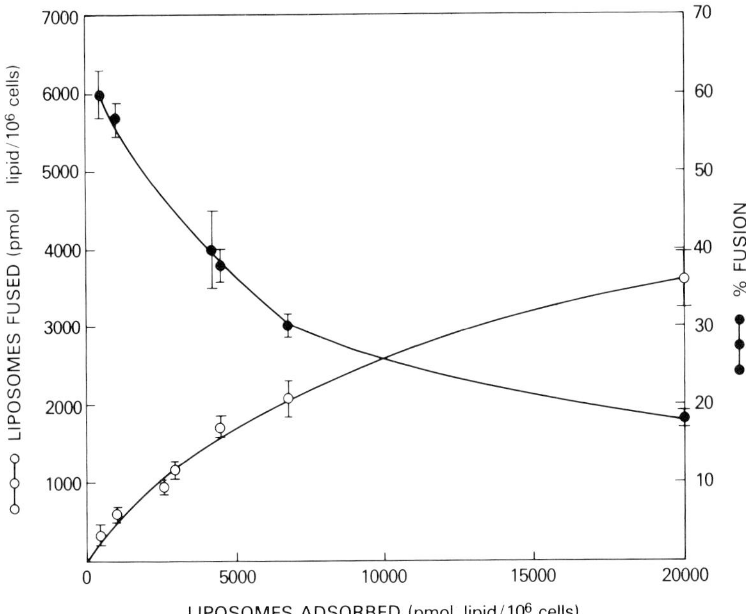

FIGURE 3.   Quantitation of liposome-cell interaction by fluorescence RET. Chromaffin cells (10^6) were incubated with liposomes (phosphatidylcholine/phosphatidylserine/cholesterol/NBD-phosphatidylethanolamine/rhodamine-phosphatidylethanol amine, 7:1:2:0.005:0.005, molar ratios) for 60 min at 37°C. The efficiency of the fluorescence energy transfer of the original liposomes and that of the washed cell suspension was analyzed in a spectrofluorimeter before and after addition of Triton® X-100 (final concentration 0.2% v/v). The extent of liposome-cell association and of liposome-membrane fusion, respectively, was calculated according to Pagano et al.[41] Open circles: amount of liposomes adsorbed to the cell surface expressed in nanomoles lipid per 10^6 cells. Closed circles: fraction of liposomes (% of those that interacted with the cells) that have actually fused with the plasma membrane.

by endocytosis. The fraction of fused liposomes is inversely proportional to the amount of liposomes adsorbed to the cells. A quantitative estimate for the standard liposomes yields that out of approximately 4×10^5 liposomes which adhere to the cells, approximately 1.25×10^5 will transfer their contents into the cytoplasm within 1 hr. In terms of the cytoplasmic concentration of the injected macromolecules, this means that their intracellular concentration will be equal to approximately 1/150 of that originally entrapped into the liposomes. This transfer efficiency can be increased by a factor of 3 to 4, using the optimized conditions, viz., cardiolipin-containing liposomes and lanthanum as additional fusogen.

III. LIPOSOME-MEDIATED MODULATION OF CHROMAFFIN CELL PHYSIOLOGY

So far in this chapter we have mainly dealt with the techniques for analyzing liposome cell fusion. As outlined before, this delivery system has been developed with the purpose to by-pass the permeability barrier and to keep the cells still intact. Structural and functional integrity of the cells have to be retained following their interaction with the liposomes. This was tested by several assays. First of all, cell viability, as judged by trypan blue exclusion, was not impaired following incubation of the cells with up to 5 μM lipid per 10^6 cells, nor did the cells release lactate dehydrogenase. Liposome-mediated catecholamine secretion was found to be truly exocytotic, as determined by the concomitant release of dopamine-β-hydroxylase. Furthermore, using standard conditions (1 μM lipid per 10^6 cells), catecholamine secretion following challenge with acetylcholine or with potassium remained essentially unchanged. Thus, liposome-cell interactions leave the cells and in particular the natural secretory machinery intact, and as shown below, are a suitable and an effective means to modulate cellular physiology from within the cells.

In discussing the potential of liposome-mediated microinjection into intact cells, I want to focus on two of the many possible applications of this technique which address issues related to stimulus-secretion coupling: (1) the effects of deliberate changes in the intracellular ionic composition and (2) the possible functional role for cytoskeletal proteins in intracellular signal transduction. In both cases we measured the catecholamine release during the incubation of the cells with liposomes ("basal release") and their subsequent response to stimulation by acetylcholine.[64] In addition, we also assayed for changes in the transmembrane electrical potential, using a fluorescence probe.[65]

A. Deliberate Changes in the Intracellular Ion Composition

The earliest concepts of neurosecretion attribute a central role to calcium in stimulus-secretion coupling.[66] Extracellular calcium seemed to be a *conditio sine qua non* for evoking receptor-mediated release from practically all types of secretory cells, e.g., from the chromaffin cells.[67-69] Subsequent studies demonstrated that secretion was generally accompanied by a rise in the intracellular calcium concentration, $[Ca^{2+}]_i$.[70-72] Indeed, all the studies using permeabilized cells agree that the minimum requirement for endocytosis seems to be the elevation of the free calcium level above approximately 1 μM.[2-6,73] However, recent research tends to assess the role of intracellular calcium more ambiguously, since, depending on the secretory cell and/or the stimulus employed, release can be evoked without an apparent increase in intracellular calcium,[74,75] whereas in other cases a rise in the cytosolic calcium level is necessary, but not sufficient, to elicit secretion.[76,77] In yet other systems, secretion is accompanied by a marked, transient drop in $[Ca]_i$.[78] Right now it appears that there is no direct correlation between $[Ca^{2+}]_i$ and secretory activity, at least not in chromaffin cells.[79]

In order to assess the requirement for elevated $[Ca^{2+}]_i$ in transmitter/mediator release, calcium-loaded liposomes have been employed to induce secretory activity in mast cells,[80] the frog neuromuscular junction,[81] rat brain synaptosomes,[82] rabbit carotid body cells,[83] and adrenal medullary slices.[84] In the latter preparation catecholamine secretion was also evoked, when the slices were incubated with liposomes loaded with sodium, but not with potassium. This finding is in line with the conventional picture of the stimulus-evoked events, which comprises rises in the level of intracellular calcium and sodium.[1,66-69,85] An additional argument for a regulatory role of elevated intracellular calcium was seen in the fact that secretion induced by increasing intracellular sodium was abolished, when the liposomes contained, in addition to sodium, the calcium chelator, EDTA.[84]

Using isolated bovine adrenal medullary chromaffin cells, we studied in detail the effects of Na^+, K^+, and Ca^{2+}-loaded liposomes on catecholamine secretion and on membrane potential. The extent of secretion was taken as a suitable parameter to optimize liposome-mediated microinjection and to assess the effects of lipid vesicles of various compositions on receptor-mediated stimulation.[18,25,56,86]

Liposomes containing Na^+ or Ca^{2+} elicited a dose-dependent increase in catecholamine output; however, the ones containing Na^+ were more effective than those containing Ca^+ (Table 2). In all cases, liposomes, containing K^+ served as an internal control, since over a wide range of liposome concentrations, an increase in $[K^+]_i$ did not measurably affect the basal catecholamine release. In addition, Na^+-loaded liposomes were also found to induce a lasting, dose-dependent depolarization, whereas introduction of K^+ and of Ca^{2+}, respectively, did not alter the transmembrane electrical potential. In a parallel study of ^{3}H-serotonin release from rat basophilic leukemia (RBL) cells, we found very similar effects for the same types of liposomes, with the one significant difference that in this system Ca^{2+}-loaded liposomes were more effective than the Na^+-containing ones.[49,87]

Table 2
**EFFECT OF LIPOSOMES CONTAINING VARIOUS CATIONS
ON CATECHOLAMINE RELEASE AND TRANSMEMBRANE
ELECTRICAL POTENTIAL[86,116]**

Liposome contents	Catecholamine release (% of control)	Membrane potential (mV)
Control, no liposomes	100	−54
K^+ (150 mM)	100	−54
K^+ (135 mM) and Ca^{2+} (5 mM)	155	−54
Na^+ (150 mM)	225	−24

Note: 10^6 isolated chromaffin cells were incubated with 2 μmol each of liposomes (phosphatidylcholine/phosphatidylserine/cholesterol, 7:1:2) for 60 min at 37°C containing the various cations. Following incubation the cells were washed twice; the first supernatant was analyzed for catecholamine contents.[64] The membrane potential of the resuspended cells was assayed fluorometrically using a potentially sensitive dye.[65]

These results shed some light on important questions related to the ionic requirements of intracellular transmission of the stimulus signal. In several secretory systems which do not express voltage-sensitive calcium channels in their plasma membranes, e.g., mast cells or RBL cells, secretion is not accomplished by nonspecific membrane depolarization, e.g., by increasing the external $[K^+]$.[88] In these cells, elevation of $[Ca^{2+}]_i$ is a necessary and a *sufficient* signal to trigger exocytosis.[89] Thus, Ca^{2+}-ionophores are potent secretagogues in mast cells or RBL.[90] In other systems, such as the chromaffin cells, however, a rise in $[Ca]_i$, mediated by ionophores, will induce only moderate levels of secretion.[78,91] In contrast, elevation of $[Na^+]_i$ in these cells might mimic some basic steps of stimulus-secretion coupling, either directly[92] or indirectly via events related to the depolarization of the plasma membrane.

The data from liposome-mediated manipulations of the cytoplasmic ion composition seem to confirm this latter line of thought. The cytosolic level of chromaffin cells that have been fused with Ca^{2+}-loaded liposomes will have increased to approximately 6 μM,[11] which is well above the experimentally observed values observed following stimulation of the cells with acetylcholine or potassium.[70-72] In permeabilized cells, elevation of the free calcium concentration to similar values was shown to elicit sizeable catecholamine secretion.[70-72,79,93,94] The level of intracellular calcium, measured in intact cells upon addition of calcium ionophores, is of similar magnitude, but was found to be an insufficient secondary messenger to effectively stimulate the secretory machinery.[78] The liposome-mediated elevation of $[Na^+]_i$, on the other hand, seems to be a more potent stimulus for several reasons: (1) it is accompanied by a depolarization of the plasma membrane, which in turn could trigger the subsequent steps in the physiological pathways leading to augmented secretion; (2) elevated $[Na^+]_i$ might enhance intracellular Na^+/Ca^{2+} exchange, as described by Gratzl in Chapter 5, thus increasing $[Ca^{2+}]_i$ and triggering secretion, in line with the conventional picture of Ca^{2+}-dependent exocytosis; (3) alternately, the effects of Na^+-containing liposomes could also be explained by the model postulated by Uvnäs and Aborg:[95] increasing $[Na^+]_i$ might result in the exchange of Na^+ for intragranular catecholamines and their subsequent secretion; (4) our results might also be viewed in the light of recent findings that an increase in $[Na^+]_i$ is a potent trigger to stimulate the PI turnover.[96] The correlation between increased PI turnover and catecholamine secretion is discussed by Schneider in Chapter 11 of this volume.

Our data support the view that in intact cells elevation of $[Ca^{2+}]_i$ is *not sufficient* to elicit

Table 3
EFFECT OF THE LIPID COMPOSITION AND INCUBATION CONDITIONS ON CELL ASSOCIATION AND CATECHOLAMINE RELEASE BY SODIUM-CONTAINING LIPOSOMES[a]

Lipid composition and incubation conditions[b]	Cell-association[c] (% vesicle fused)	Catecholamine release[d] (% of control)
PC/PS/Chol, 7:1:2	35	100
PC/PS/Chol, 3:5:2	11	8
PC/PS, 1:1	5	10
PC/Chol, 1:1	23	100
DMPC/PS/Chol, 7:1:2	N.D.	95
PC/CAR/Chol, 7:1:2	33	150
PC/CAR/Chol, 2:3:5	22	250
PC/PE/Chol, 6:2:2	24	80
PC/PA/Chol, 7:1:2	N.D.	25
PC/PS/Chol 7:1:2 + 0.5% BSA	60	50
PC/PS/Chol 7:1:2 + 100 μM La^{3+}	50	150
PC/CAR/Chol 2:3:5 + 100 μM La^{3+}	25	380

[a]   Data summarized from References 18, 25, 86, and unpublished results.

[b]   Abbreviations for the various lipids are CAR, cardiolipin; Chol, cholesterol; DMPC, dimyristoylphosphatidylcholine; PA, phosphatidic acid; PC, egg phosphatidylcholine; PE, phosphatidylethanolamine; PS, phosphatidylserine. In all experiments, 10^6 cells were incubated with liposomes (1 μM lipid) for 60 min at 37°C in buffer 1 + 2.2 mM Ca^{2+}.

[c]   Extent of cell association/fusion was determined by fluorescence resonance energy transfer. All liposomes (except those containing PA) were fluorescently labeled with NBD-PE and rhodamine-PE. The data show the percentage of cell-adsorbed vesicles that have actually fused with the cell membranes; see text for details.

[d]   Following incubation of the cells for 1 hr with the various liposomes, catecholamines in the supernatant were assayed according to Lelkes et al.[64] The data are expressed as % of the release evoked by the standard Na$^+$-containing liposomes (PC/PS/Chol, 7:1:2).

sizeable secretion, but that rather a rise in the intracellular calcium is probably necessary, together with the concomitant activation of yet unidentified second (third, nth?) messenger(s), e.g., Na$^+$, PI metabolites, nucleotides, changes in pH$_i$ and/or in the assembly of cytoskeletal proteins, etc., to account for the intracellular aspects of stimulus transduction.

The efficacy of Na$^+$-loaded liposomes to modulate catecholamine release was found to depend largely on the lipid composition of the carrier vesicles[18,25,56,86] (Table 3). As detailed earlier, the efficiency of liposome-cell interactions is partially determined by the formulation of the vesicles; however, liposomes containing phosphatidylethanolamine (PE) were bound to the cell surface and fused equally well as the standard vesicles containing phosphatidylserine, and yet the extent of induced catecholamine release decreased with increasing the PE concentration.[18,25] Similarly, Na$^+$-loaded liposomes that contained a highly fusogenic lipid, phosphatidic acid, in the lipid mixture were found to elicit only a poor catecholamine secretion.[86] In contrast, changing the fatty acid chain length or the degree of saturation of the major liposome constituent, phosphatidylcholine, had no measurable effect on the amount of catecholamine release induced by Na$^+$-loaded liposomes.[25]

Thus, besides providing access to the cytoplasm, liposomes seem to be an interesting tool for modulating cellular functions related to the lipid composition of the plasma membrane. As pointed out by Rosenheck (Chapter 10) the physical state of the plasma membrane lipids is modified upon stimulation of the secretory apparatus.[97] The inverse experiments, viz., to study stimulus-secretion coupling following deliberate changes in the membrane lipid composition, could provide new information on lipid/protein interactions, regulating receptor and/or channel functions.

B. Alterations in the State of Assembly of Microfilaments

Ever since the discovery of contractile proteins in nonmuscle cells, in particular, in secretory systems, a functional role has been proposed for the participation of cytoskeletal proteins in stimulus-secretion coupling.[99-102] The highly organized microtrabecular lattice in the cytoplasm of adrenal chromaffin cells poses a steric constraint to random movement of secretory granules.[103] Similarly, ordered arrays of microtubules at the cytoplasmic surface of presynaptic membranes suggest that the size of these lattices might determine the quantal unit of neurotransmitter release.[104] Therefore, modulations in the state of assembly of these filamentous barriers could be expected to facilitate exocytosis. Indeed, in neutrophils, lysosomal enzyme secretion is augmented upon treating the cells with cytochalasin B, a drug which disrupts microfilaments.[105] However, in other systems it was found that drugs directed against either the microfilaments or the microtubules inhibited neurosecretion from sympathetic nerves[106] and also, catecholamine release from chromaffin cells.[107] Further experiments revealed that vinca alkaloids, which are potent antimicrotubular drugs, selectively inhibit receptor-mediated $^{45}Ca^{2+}$ uptake and catecholamine release from chromaffin cells, while secretion evoked by potassium was not impaired.[108] By now it is clear that a number of the "specific" anticytoskeletal drugs exhibits a wide variety of nonspecific effects, e.g., directed toward the nicotinic receptor.[108,109] Thus, great caution should be exercised when implicating biological mechanism from inhibitory effects of drugs that might have proven specific in the test tube, but less so in complex systems, such as cells or tissues. This general truth has been proven for a variety of "specific" compounds, such as antagonists to calmodulin,[110] for α_2-adrenergic blockers,[111] for oligopeptide inhibitors of metalloprotease activity,[112] and as mentioned, for anticytoskeletal drugs.

Thus, investigations into the functional role of cytoplasmic determinants, such as cytoskeletal proteins in stimulus-secretion coupling, require direct access to the cytoplasmic compartment. In recent years this goal has been achieved by various physical and chemical techniques to permeabilize plasma membranes, either by exposure to high electrical fields,[73] by treatment with low concentrations of detergents,[3-5] or pore-forming toxins.[6] In cells permeabilized by electric pulses or by toxins, access to the cytoplasm is limited to small molecules only, while detergent-permeabilized cells can accommodate larger molecules, e.g., immunoglobulins. In studying the cytoplasmic determinants of stimulus-secretion coupling, these permeabilized chromaffin cells exhibit a major disadvantage: catecholamine release from these cells cannot be activated via nicotinic receptor agonists or any other secretagogues affecting transmembrane ion fluxes. However, since exposure of these cells to micromolar concentrations of calcium is sufficient to generate catecholamine release, they can be valuable tools in exploring some of the cytoplasmic factors, which might modulate Ca^{2+}-induced exocytotic membrane fusion.

Among the numerous compounds tested in electrically permeabilized chromaffin cells were anticytoskeletal drugs such as cytochalasin B.[73] The experimental evidence in those "leaky" cells unequivocally argues against a direct involvement of microfilaments in Ca^{2+}-triggered catecholamine release. In a related set of experiments, Whitaker and Baker[113] studied calcium-dependent exocytosis in vitro, in a secretory granule/plasma membrane preparation from sea urchin eggs. A number of anticytoskeletal drugs, including phalloidin, colchicine, vinblastine, as well as antibodies to actin, was found to be without effect on exocytosis, suggesting that neither microtubules nor microfilaments participate intimately in exocytosis in this model system.

More recent studies, however, seem to indicate that Ca^{2+}-induced catecholamine secretion from *digitonin-permeabilized* chromaffin cells is modulated by anticytoskeletal drugs[114] and/or upon addition of microfilamental proteins, such as actin or heavy meromyosin.[115] These data highlight some of the discrepancies between permeabilized cells and other model systems for exocytosis; however, they do not conclusively assign *a* or *the* function for cytoskeletal proteins during stimulus-secretion coupling. Baker and Knight[73] pointed out that none of these models for exocytosis permits the investigation of alternate functional roles for the cytoskeleton, e.g., related to signal transduction across the plasma membrane.

Such a possible involvement of microfilaments in the intracellular processing of the stimulus signal was the working hypothesis that led us to search for alternate ways to access the cytoplasm, while leaving the cells functionally intact.

First evidence for such an involvement of microfilaments in stimulus-secretion coupling was obtained following liposome-mediated introduction of DNAse I into isolated chromaffin cells.[116] Upon incubation with DNAse I-loaded liposomes, it was found that the cells released significantly more catecholamines than nontreated control cells or cells fused with potassium-containing liposomes. In addition, the plasma membrane was depolarized from -55 to -34 mV. In contrast to an increase in the level of "basal" release, incubation of the cells with DNAse I-loaded liposomes did not alter their secretory response to subsequent stimulation with acetylcholine.

DNAse I strongly binds to G-actin causing a shift in the G-F equilibrium of actin.[117] Therefore, introduction of DNAse I is expected to cause a depletion of the free cytosolic G-actin pool available for polymerization and eventually result in the disappearance of F-actin, a sizeable portion of which is associated with the plasma membranes of the isolated chromaffin cells.[118] The effect of DNAse I on basal catecholamine secretion can be viewed mechanistically as the removal of the submembranal web of the actin filaments. This network of microfilaments has been postulated to control the level of basic secretion by obstructing massive, random exocytosis under resting conditions.[116,119] Thus, the observed DNAse I effects seem to mimic certain stages in stimulus transduction, where fusion of the storage granules with the plasma membranes is preceded by solation of the submembranal web of microfilaments.[120]

The increase in catecholamine output of DNAse I-treated cells was found to be accompanied by a significant depolarization of the plasma membranes, indicating a possible link between the state of microfilaments and transmembranal ion fluxes. Recent more-detailed studies have confirmed this hypothesis.[121] Introduction of DNAse I and of heavy meromyosin (HMM) caused a significant increase of $^{45}Ca^{2+}$ influx into chromaffin cells, which was not found in control cells treated with buffer liposomes or liposomes containing N-ethylmaleimide-poisoned HMM. The data suggest an interaction between the microfilaments and the voltage-activated Ca^{2+} channels, since the augmented uptake of $^{45}Ca^{2+}$ could be blocked by cobalt ions.

To further investigate the possibility for a mutual regulation of microfilament assembly, transmembranal ion fluxes, and catecholamine release, Friedman and co-workers[122] introduced several soluble fragments of myosin into the cytoplasm of isolated chromaffin cells: heavy meromyosin, which is fully active mechanochemically, and the myosin subfragment 1 (S_1), which contains the ATPase activity, but is mechanochemically less competent than HMM. As an internal control for the technique and for the interpretation of the data, cells were also incubated with liposomes containing these macromolecules after irreversible inactivation by N-ethylmaleimide (NEM).

Treatment of the cells with HMM and S_1 resulted in augmented "basal" catecholamine release during the incubation period concomitant with a sustained membrane depolarization (Table 4). HMM induced more release and also caused a larger depolarization than S_1. No change in the basal catecholamine output was observed when the cells were treated with the NEM-poisoned myosin fragments. The resting membrane potential was not changed by NEM-S_1, while NEM-HMM caused a slight hyperpolarization. Subsequent stimulation with acetylcholine resulted in augmented secretion for both HMM and S_1-treated cells, while the response to the secretagogue was significantly decreased following treatment of the cells with NEM-HMM.

Table 4
EFFECT OF LIPOSOME-MEDIATED
INTRODUCTION OF CYTOSKELETON-SPECIFIC
MACROMOLECULES ON CATECHOLAMINE
SECRETION AND MEMBRANE POTENTIAL IN
ISOLATED CHROMAFFIN CELLS[a]

Liposome contents[b]	Catecholamine release[c] (% of control)	Membrane potential[d] (mV)
Control, K$^+$-liposomes	100	−54
DNAse I	200	−31
HMM	190	−16
S-1	120	−35
NEM-HMM	75	−60

[a] Data summarized from References 116, 119, and 122.
[b] The macromolecules (5 mg/mℓ) were incorporated into French pressed liposomes, composed of PS/PC/Chol (7:1:2). The purified liposomes were incubated at 1 μm lipid per 10^6 cells for 60 min at 37°C in buffer 1 + 2.2 mM Ca^{2+}. For details and abbreviations, see text.
[c] Upon termination of the experiment, catecholamines in the incubation medium were analyzed according to Lelkes et al.[64] The data are expressed as % of control cells, which were processed with K$^+$-liposomes in the incubation medium.
[d] The membrane potential of the washed cell suspension was determined fluorometrically using a potential sensitive dye, according to Friedman et al.[65]

Friedman and co-workers[122] also reported that HMM, but not NEM-HMM, caused an increase in ^{22}Na$^+$ uptake into chromaffin cells, without effecting the rates of ^{86}Rb retention, under resting conditions and upon stimulation with acetylcholine. The stimulatory effects of HMM on both catecholamine release and ^{22}Na$^+$ uptake were not inhibited by blockers of the acetylcholine receptor-associated ion channel and of the voltage-sensitive Ca$^+$ channels, respectively. However, HMM-induced catecholamine secretion was suppressed upon removal of extracellular Na$^+$, while the uptake of ^{22}Na$^+$ was greatly reduced in the presence of the inhibitor of the Na$^+$/H$^+$ antiporter, amiloride.

In summary, these data obtained in *intact* chromaffin cells treated with cytoskeletal proteins or related macromolecules provide, for the first time, clear evidence for an involvement of microfilaments in the control of stimulus-secretion coupling. The functional role of microfilaments, according to these studies, is believed to be a dual one. The submembranal web might serve as a physical barrier limiting massive, random contact of granules with the membrane. Solation of F-actin in the vicinity of the plasma membrane, e.g., evoked by elevated [Ca]$_i$ following stimulation, could be a prerequisite for effective stimulus-secretion coupling. Indeed, Perrin and Aunis[123] recently reported such a localized reorganization of an actin-binding protein, α-fodrin, in chromaffin cells. Such changes in the state of assembly of microfilaments following receptor-mediated stimulation have also been observed in other secretory cells, such as human neutrophils,[124,125] rat basophilic leukemia cells,[126] as well as in chromaffin cells.[86,118,127] In all these cells, secretion was accompanied by a transient solation of cellular F-actin. This concept has recently been summarized by Pollard and co-workers[120] and presented as a working hypothesis for the role of calcium-induced gel-sol transitions of cytoplasmic actin, which could provide a convenient control of the intracellular mobility to storage granules.

In addition, the HMM-related studies of Friedman and co-workers[122] suggest that microfilaments might actively participate in intracellular signal transduction by establishing a (physical?) contact to the Na^+/H^+ antiporter and/or other ion channels. Besides circumstantial evidence for a link between microfilaments and the voltage-sensitive calcium channels,[116,121] a possible scenario for such an interaction is the following: upon stimulation of the cells, ATP is consumed, e.g., during phosphorylation of the myosin light chain. ATP hydrolysis might lead to a localized generation of protons, which in turn will activate the Na^+/H^+ antiporter. These concepts are in line with recent observations[128] that changes in the cytosolic pH may be a general scheme, common to many cells, for intracellular signal transduction, in particular, in stimulus-secretion coupling.

IV. CONCLUDING REMARKS

Liposome-mediated delivery of macromolecules is certainly a feasible and attractive alternative to study cytoplasmic determinants of stimulus-secretion coupling in *intact* cells. The technical details outlined for the chromaffin cells might have to be modified when working with other cellular systems. In particular, it might be desirable to develop a liposomal vector for cytoplasmic delivery suitable for targeting to cells maintained in prolonged tissue culture.

The importance of employing *intact* cells for the study of cytoplasmic determinants of receptor-mediated cellular events is clearly emphasized by the results, which were not observed previously using electrically or chemically permeabilized cells. In particular, the use of this technique has provided unequivocal evidence that at least one class of the cytoskeletal proteins, the microfilaments, is involved in the intracellular signal transmission and regulation of secretion in the adrenal chromaffin cells. We can expect that liposomal vectors and/or other alternate means will continue to be employed to deliberately modulate the state of assembly of the cytoskeleton in *intact* cells. Such studies will certainly provide a more detailed functional assignment for the various classes of cytoskeletal proteins (microtubules, microfilaments, and the intermediate filaments). It will be of particular importance to apply these techniques to unravel the functional role for the increasing number of those other cytoplasmic proteins that are known to interact with the cytoskeleton.

ACKNOWLEDGMENTS

Part of the original work described here was carried out at the Department of Membrane Research, The Weizmann Institute of Science, Rehovoth, Israel, and supported by a grant-in-aid from the Israel Academy of Science and Humanities. I am indebted to my collaborators, J. E. Friedman, K. Rosenheck, A. Oplatka, and R. Dardik, for their contributions to an exciting joint venture. The manuscript for this chapter would have never been typed without the Superwriter® wordprocessing program.

REFERENCES

1. **Pollard, H. B., Ornberg, R., Levine, M., Kelner, K., Morita, K., Levine, R., Forsberg, E., Brocklehurst, K., Duong, L., Lelkes, P. I., Heldman, E., and Youdim, M. B. Y.,** Hormone secretion by exocytosis with emphasis on information from the chromaffin cell system; view from the vantage point of the chromaffin cells, in *Vitamins and Hormones,* Vol. 42, Auerbach, G., Ed., Academic Press, Orlando, Fla., 1986, 109.

2. **Baker, P. F. and Knight, D. E.,** Calcium-dependent exocytosis in bovine adrenal medullary cells with leaky plasma membranes, *Nature (London),* 276, 620, 1978.

3. **Brooks, J. C. and Treml, S.,** Catecholamine secretion by chemically skinned cultured chromaffin cells, *J. Neurochem.,* 40, 468, 1983.
4. **Dunn, L. A. and Holz, R. W.,** Catecholamine secretion from digitonin-treated adrenal medullary chromaffin cells, *J. Biol. Chem.,* 258, 4989, 1983.
5. **Wilson, S. P. and Kirshner, N.,** Calcium-evoked secretion from digitonin-permeabilized adrenal medullary chromaffin cells, *J. Biol. Chem.,* 258, 4994, 1983.
6. **Ahnert-Hilger, G., Bhahkdit, S., and Gratzl, M.,** Minimal requirements for exocytosis. A study using PC 12 cells permeabilized with staphylococcal alpha-toxin, *J. Biol. Chem.,* 260, 12730, 1985.
7. **Keith, C., DiPaola, M., Maxfield, F. R., and Shelanski, M. M.,** Microinjection of Ca^{++}-calmodulin causes a localization of depolymerization of microtubules, *J. Cell Biol.,* 97, 1918, 1983.
8. **Fernandez, J. M., Neher, E., and Gomperts, B. D.,** Capacitance measurements reveal stepwise fusion events in degranulating mast cell, *Nature (London),* 312, 453, 1984.
9. **Doxesey, S. J., Sambrook, J., Helenius, A., and White, J.,** An efficient method for introducing macromolecules into living cells, *J. Cell Biol.,* 101, 19, 1985.
10. **Schlegel, R. A. and Rechtsteiner, M. C.,** Erythrocyte mediated transfer: applications, in *Microinjection and Organelle Transfer Techniques: Methods and Applications,* Celis, J. E., Graesmann, A., and Loyter, A., Eds., Academic Press, New York, 1985.
11. **Lelkes, P. I. and Friedman, J. E.,** Interaction of French-pressed liposomes with isolated bovine adrenal chromaffin cells. Characterization of the cell-liposome interactions, *J. Biol. Chem.,* 260, 1796, 1985.
12. **Kenigsberg, R. L. and Trifaró, J. M.,** Microinjection of calmodulin antibodies into cultured chromaffin cells blocks catecholamine release in response to stimulation, *Neuroscience,* 14, 335, 1985.
13. **Kenigsberg, R. L. and Trifaró, J. M.,** A technique for the microinjection of macromolecules into viable chromaffin cells in culture, *J. Neurosci. Methods,* 13, 103, 1985.
14. **Poste, G.,** The interaction of lipid vesicles (liposomes) with cultured cells and their use as carriers for drugs and macromolecules, in *Liposomes in Biological Systems,* Gregoriadis, G. and Allison, A. C., Eds., Wiley-Interscience, New York, 1980, 102.
15. **Szoka, F., Jr. and Papahadjopoulos, D.,** Comparative properties and methods of preparations of lipid vesicles (liposomes), *Ann. Rev. Biophys. Bioeng.,* 9, 467, 1980.
16. **Barenholzt, Y., Amselem, S., and Lichtenberg, D.,** A new method for preparation of phospholipid vesicles (liposomes). French press, *FEBS Lett.,* 99, 210, 1979.
17. **Hamilton, R. L., Goerke, J., Guo, L. S. S., Williams, M. C., and Havel, R. J.,** Unilammelar liposomes made with the French pressure cell: a simple preparative and semiquantitative technique, *J. Lipid Res.,* 21, 981, 1980.
18. **Lelkes, P. I.,** The use of French pressed vesicles for efficient incorporation of bioactive macromolecules and as drug carriers in vitro and in vivo, in *Liposome Technology,* Vol. 1, Gregoriadis, G., Ed., CRC Press, Boca Raton, Fla., 1983, 51.
19. **Lelkes, P. I. and Tandeter, H. B.,** Studies on the methodology of the carboxyfluorescein assay and on the mechanism of liposome stabilization by red blood cells *in vitro, Biochim. Biophys. Acta,* 716, 410, 1982.
20. **Straubinger, R. M., Hong, K., Friend, D. S., and Papahadjopoulos, D.,** Endocytosis of liposomes and intracellular fate of encapsulated molecules: encounters with a low pH compartment after internalization in coated vesicles, *Cell,* 32, 1069, 1983.
21. **Leserman, L. D., Weinstein, J. N., Blumenthal, R., and Terry, W. D.,** Receptor-mediated endocytosis of antibody-opsonized liposomes by tumor cells, *Proc. Natl. Acad. Sci. U.S.A.,* 77, 4089, 1980.
22. **Nir, S., Bentz, J., Wilschut, J., and Duzgunes, N.,** Aggregation and fusion of phospholipid vesicles, *Prog. Surf. Sci.,* 13, 1, 1983.
23. **Papahadjopoulos, D.,** Calcium induced phase changes and fusion in natural and model membranes, in *Cell Surface Reviews,* Vol. 5, Poste, G. and Nicolson, G., Eds., North-Holland, Amsterdam, 1978, 765.
24. **Cohen, F. S., Akabas, M. H., Zimmerberg, J., and Finkelstein, A.,** Parameters affecting the fusion of unilamellar phospholipid vesicles with planar bilayer membranes, *J. Cell Biol.,* 98, 1054, 1984.
25. **Dardik, R.,** Optimalization of Liposome-Chromaffin Cell Interactions, M. Sc. thesis, The Feinberg Graduate School, Weizmann Institute of Science, Rehovoth, Israel, 1984.
26. **Akabas, M. H., Cohen, F. S., and Finkelstein, A.,** Separation of the osmotically driven fusion event from vesicle-planar membrane attachment in a model system for exocytosis, *J. Cell Biol.,* 98, 1063, 1984.
27. **Steinkamp, J.,** Flow cytometry, *Rev. Sci. Instrum.,* 55, 1375, 1984.
28. **Blumenthal, R., Weinstein, J. N., Sharrow, S. O., and Henkart, P.,** Liposome-lymphocyte interactions: saturable sites for transfer and intracellular release of liposome contents, *Proc. Natl. Acad. Sci. U.S.A.,* 74, 5603, 1977.
29. **Lelkes, P. I., Klein, L., Marikovsky, Y., and Eisenbach, M.,** Liposome-mediated transfer of macromolecules into flagellated cell envelopes from bacteria, *Biochemistry,* 23, 563, 1984.
30. **Struck, D. K., Hoekstra, D., and Pagano, R. E.,** Use of resonance energy transfer to monitor membrane fusion, *Biochemistry,* 20, 4093, 1981.

31. **Kercret, H., Chiovetti, R., Jr., Fountain, M. W., and Segrest, J. P.,** Plasma membrane mediated-leakage of liposomes induced by interaction with murine thymocytic leukemia cells, *Biochim. Biophys. Acta,* 733, 65, 1983.

32. **Senior, J. and Gregoriadis, G.,** Methodology in assessing liposomal stability in the presence of blood, clearance from the circulation of injected animals, and uptake by tissues, in *Liposome Technology,* Vol. 3, Gregoriadis, G., Ed., CRC Press, Boca Raton, Fla., 1984, 263.

33. **Scherphof, G. L., Damen, J., and Wilschut, J.,** Interactions of liposomes with plasma proteins, in *Liposome Technology,* Vol. 3, Gregoriadis, G., Ed., CRC Press, Boca Raton, Fla., 1984, 205.

34. **Van Renswoude, J. and Hoekstra, D.,** Cell induced leakage of liposome contents, *Biochemistry,* 20, 540, 1981.

35. **Weinstein, J. N., Ralston, E., Leserman, L. D., Klausner, R. D., Dragsten, P., Henkart, P., and Blumenthal, R.,** Self quenching of carboxyfluorescein fluorescence: use in studying liposome stability and liposome-cell interaction, in *Liposome Technology,* Vol. 3, Gregoriadis, G., Ed., CRC Press, Boca Raton, Fla., 1984, 183.

36. **Lelkes, P. I.,** Methodological aspects dealing with stability measurements of liposomes *in vitro* using the carboxyfluorescein assay, in *Liposome Technology,* Vol. 3, Gregoriadis, G., Ed., CRC Press, Boca Raton, Fla., 1984, 225.

37. **Szoka, F. C., Jacobson, K., and Papahadjopoulos, D.,** Use of aqueous space markers to determine the mechanism of interaction between phospholipid vesicles and cells, *Biochim. Biophys. Acta,* 551, 295, 1979.

38. **Allen, M.,** Calcein as a tool in liposome methodology, in *Liposome Technology,* Vol. 3, Gregoriadis, G., Ed., CRC Press, Boca Raton, Fla., 1984, 172.

39. **Gibson, G. A. and Loew, L. M.,** Phospholipid vesicle fusion monitored by fluorescence energy transfer, *Biochem. Biophys. Res. Commun.,* 88, 135, 1979.

40. **Van der Werf, P. and Ullman, E. F.,** Monitoring of phospholipid vesicle fusion by fluorescence energy-transfer between membrane bound labels, *Biochim. Biophys. Acta,* 596, 302, 1980.

41. **Pagano, R. E., Schroit, A. J., and Struck, D. K.,** Interactions of phospholipid vesicles with mammalian cells *in vitro*: studies of mechanism, in *Liposomes: From Physical Structure to Therapeutic Applications,* Vol. 7, Knight, C. G., Ed., Elsevier/North-Holland, Amsterdam, 1981, 323.

42. **Hoekstra, D.,** Role of lipid phase separations and membrane hydration in phospholipid vesicle fusion, *Biochemistry,* 21, 2833, 1982.

43. **Nichols, J. W. and Pagano, R. E.,** Use of resonance energy transfer to study the kinetics of amphiphile transfer between vesicles, *Biochemistry,* 21, 1720, 1982.

44. **Keller, P. M., Person, S., and Snipes, W.,** A fluorescence enhancement assay of cell fusion, *J. Cell Sci.,* 28, 167, 1977.

45. **Hoekstra, D., De Boer, T., Klappe, K., and Wilschut, J.,** Fluorescence method for measuring the kinetics of fusion between biological membranes, *Biochemistry,* 23, 5675, 1984.

46. **Axelrod, D., Koppel, D. E., Schlessinger, J., Elson, E., and Webb, W. W.,** Mobility measurement by analysis of fluorescence photobleaching recovery kinetics, *Biophys. J.,* 15, 1055, 1976.

47. **Schneider, A. S., Herz, R., and Rosenheck, K.,** Stimulus-secretion coupling in chromaffin cells isolated from bovine adrenal medulla, *Proc. Natl. Acad. Sci. U.S.A.,* 74, 5036, 1977.

48. **Ballentine, R. and Burford, D. D.,** Differential density separation of cellular suspensions, *Anal. Biochem.,* 1, 263, 1960.

49. **Lelkes, P. I. and Sagi-Eisenberg, R.,** Interaction of liposomes with rat basophilic leukemia cell: modulation of degranulation and of transmembrane potential, submitted.

50. **Martin, T. W. and Lagunoff, D.,** Activation of histamine secretion from rat mast cells by aqueous dispersions of phosphatidylserine, *Biochemistry,* 19, 3106, 1980.

51. **Pagano, R. E. and Takeichi, M.,** Adhesion of phospholipid vesicles to Chinese hamster fibroblasts. Role of cell surface proteins, *J. Cell Biol.,* 74, 531, 1977.

52. **Van Renswoude, A. J. B. M., Westenberg, P., and Scherphof, G.,** In vitro interaction of Zajdela ascites hepatoma cells with lipid vesicles, *Biochim. Biophys. Acta,* 558, 22, 1979.

53. **Lelkes, P. I.,** unpublished data, 1985.

54. **Pagano, R. E. and Weinstein, J. N.,** Interactions of liposomes with mammalian cells, *Ann. Rev. Biophys. Bioeng.,* 7, 435, 1978.

55. **Huang, L.,** Liposome-cell interactions in vitro, in *Liposomes,* Ostro, M., Ed., Marcel Dekker, New York, 1983, 87.

56. **Dardik, R., Rosenheck, K., and Lelkes, P. I.,** Optimization of liposome interaction with bovine adrenal medullary chromaffin cells, manuscript submitted.

57. **Hammoudah, M. M., Nir, S., Bentz, J., Mayhew, E., Stewart, T. P., Hui, S. W., and Kurland, R. J.,** Interactions of La^{3+} with phosphatidylserine vesicles: binding, phase transition, leakage, ^{31}P-NMR and fusion, *Biochim. Biophys. Acta,* 645, 102, 1981.

58. **Bental, M., Lelkes, P. I., Scholma, J., Hoekstra, D., and Wilschut, J.,** Ca^{++} independent, protein-mediated fusion of chromaffin granule ghosts with liposomes, *Biochim. Biophys. Acta,* 774, 296, 1984.

59. **Wilschut, J., Düzgünes, N., Fraley, R., and Papahadjopoulos, D.,** Studies on the mechanism of membrane fusion: kinetics of calcium ion induced fusion of phosphatidylserine vesicles, *Biochemistry,* 19, 6011, 1980.

60. **Düzgünes, N., Wilschut, J., Fraley, R., and Papahadjopoulos, D.,** Studies on the mechanism of membrane fusion. Role of head-group composition in calcium- and magnesium-induced fusion of mixed phospholipid vesicles, *Biochim. Biophys. Acta,* 642, 182, 1981.

61. **Larrabee, A. L.,** Time-dependent changes in the size distribution of distearoylphosphatidylcholine vesicles, *Biochemistry,* 18, 3321, 1979.

62. **Wilschut, J., Holsappel, M., and Jansen, R.,** Ca^{2+}-induced fusion of cardiolipin/phosphatidylcholine vesicles monitored by mixing of aqueous contents, *Biochim. Biophys. Acta,* 690, 297, 1982.

63. **Stegman, T., Hoekstra, D., Scherphof, G., and Wilschut, J.,** Fusion of influenza virus with cardiolipin liposomes at low pH, *Biochemistry,* 24, 3107, 1985.

64. **Lelkes, P. I., Friedman, J., and Rosenheck, K.,** Direct fluorometric assay of catecholamine secretion from isolated bovine adrenal chromaffin cells, *J. Neurosci. Methods,* 13, 249, 1985.

65. **Friedman, J. E., Lelkes, P. I., Lavie, E., Rosenheck, K., Schneeweiss, F., and Schneider, A. S.,** Membrane potential and catecholamine secretion by bovine adrenal chromaffin cells: use of tetraphenyl-phosphonium distribution and carbocyanine dye fluorescence, *J. Neurochem.,* 44, 1391, 1985.

66. **Douglas, W. W.,** Stimulus-secretion coupling: the concept and clues from chromaffin and other cells, *Br. J. Pharmacol.,* 34, 451, 1968.

67. **Douglas, W. W.,** Secretomotor control of adrenal medullary secretion: synaptic, membrane and ionic events in stimulus-secretion coupling, in *Handbook of Physiology,* Sect. 7, Vol. 6, Blascko, H., Sayers, G., and Smith, A. D., Eds., American Physiological Society, Washington, D.C., 1975, 367.

68. **Douglas, W. W.,** Stimulus-secretion coupling: variations on the theme of calcium activated exocytosis involving cellular and extracellular sources of calcium, in *CIBA Foundation Symp. 54 (New Series): Respiratory Tract Mucus,* Elsevier/North-Holland, Amsterdam, 1978, 61.

69. **Rubin, R. P.,** *Calcium and Cellular Secretion,* 2nd ed., Plenum Press, New York, 1982.

70. **Knight, D. E. and Kesteven, N. T.,** Evoked transient intracellular free Ca^{2+} changes and secretion in isolated bovine adrenal medullary cells, *Proc. R. Soc. London Ser. B,* 218, 177, 1983.

71. **Burgoyne, R. D.,** The relationship between secretion and intracellular free calcium in bovine adrenal chromaffin cells, *Biosci. Rep.,* 4, 605, 1984.

72. **Kao, L.-S. and Schneider, A. S.,** Quin 2 fluorescence measurements of cytosolic calcium during stimulus-secretion coupling in adrenal chromaffin cells, *Fed. Proc.,* 43, 770, 1984.

73. **Knight, D. E. and Baker, P. F.,** Calcium-dependence of catecholamine release from bovine adrenal medullary cells after exposure to intense electrical fields, *J. Membrane Biol.,* 68, 107, 1982.

74. **Rink, T. J., Sanchez, A., and Halla, T. J.,** Diacylglycerol and phorbol ester stimulate secretion without raising cytoplasmic free calcium in human platelets, *Nature (London),* 305, 317, 1983.

75. **Pozzan, T., Gatti, A., Dozio, N., Vincenti, L. M., and Meldolesi, J.,** Calcium dependent and independent release of neurotransmitter from PC 12 cells: a role for activation of protein kinase C, *J. Cell Biol.,* 99, 628, 1984.

76. **Kao, L.-S. and Schneider, A. S.,** Muscarinic receptors on bovine chromaffin cells mediate a rise in cytosolic calcium that is independent of extracellular calcium, *J. Biol. Chem.,* 260, 2019, 1985.

77. **Korchak, H. M., Vienne, K., Rutherford, L. E., Wilkenfeldt, C., Finkelstein, M. C., and Weissmann, G.,** Stimulus response coupling in the human neutrophil. II. Temporal analysis of changes in cytosolic calcium and calcium efflux, *J. Biol. Chem.,* 259, 4076, 1984.

78. **Shoback, D. M. and Brown, E. M.,** PTH release stimulated by high extracellular potassium is associated with a decrease in cytosoloc calcium in bovine parathyroid cells, *Biochem. Biophys. Res. Commun.,* 123, 684, 1984.

79. **Heldman, E., Zimlichman, R., Keiser, H., and Pollard, H. B.,** Kinetics of secretion of catecholamines from cultured chromaffin cells: correlations with calcium influx and changes in intracellular calcium concentration, manuscript submitted.

80. **Theoharides, T. C. and Douglas, W. W.,** Secretion in mast cells induced by calcium entrapped within phospholipid vesicles, *Science,* 201, 1143, 1978.

81. **Bernon, R., Leitner, L.-M., Roumy, M., and Verna, A.,** Effect of ion-containing liposomes upon the chemoafferent activity of the rabbit carotid body superfused in vitro, *Neurosci. Lett.,* 35, 289, 1983.

82. **Rahamimoff, R., Meiri, H., Erulkar, B., and Barenholz, Y.,** Changes in the transmitter release induced by ion-containing liposomes, *Proc. Natl. Acad. Sci. U.S.A.,* 75, 5214, 1978.

83. **Crosland, R. D., Martinn, J. V., and McClure, W. O.,** Effects of liposomes containing various divalent cations on the release of acetylcholine from synaptosomes, *J. Neurochem.,* 40, 681, 1983.

84. **Gutman, Y., Lichtenberg, D., Cohen, J., and Boonyaviroj, R.,** Increased catecholamine release from adrenal medulla by liposomes loaded with sodium or calcium ions, *Biochem. Pharmacol.,* 28, 1209, 1979.

85. **Ungar, A. and Phillips, J. H.,** Regulation of the adrenal medulla, *Physiol. Rev.,* 63, 787, 1983.

86. **Friedman, J. E.,** The Possible Implications of Mechanochemical Proteins in Excitation-Secretion Coupling, M.Sc. thesis, The Feinberg Graduate School, Weizmann Institute of Science, Rehovoth, Israel, 1980.

87. **Lelkes, P. I., Sagi-Eisenberg, R., and Pecht, I.,** Fusion of liposomes with rat basophils: modulation of histamine release by internalization of vesicle contents, *Isr. J. Med. Sci.,* 18a, 13, 1982.

88. **Sagi-Eisenberg, R. and Pecht, I.,** Membrane potential changes during IgE-mediated histamine release from rat basophilic leukemia cells, *J. Membr. Biol.,* 75, 97, 1983.

89. **Parker, W. L. and Martz, E.,** Calcium ionophore A23187 as a secretagogue for rat mast cells: does it bypass inhibition by calcium flux blockers?, *Agents Actions,* 12, 276, 1982.

90. **Fewtrell, C., Lagunoff, D., and Metzger, H.,** Secretion from rat basophilic leukemia cells induced by calcium ionophores. Effect of pH and metabolic inhibition, *Biochim. Biophys. Acta,* 644, 363, 1981.

91. **Brocklehurst, K., Morita, K., and Pollard, H. B.,** Characterization of protein kinase C and its role in catecholamine secretion from bovine adrenal-medullary cells, *Biochem. J.,* 228, 35, 1985.

92. **Sorimachi, M., Nishimura, S., and Yamagami, K.,** Possible occurrence of Na^+-dependent Ca^{2+} influx mechanism in isolated bovine chromaffin cells, *Brain Res.,* 208, 442, 1981.

93. **Kilpatrick, D. L., Slepetis, R. J., Corcoran, J. J., and Kirshner, N.,** Calcium uptake and catecholamine secretion by cultured bovine adrenal medulla cells, *J. Neurochem.,* 38, 427, 1982.

94. **Holz, R. W., Senter, R. A., and Fry, R. A.,** Relationship between calcium uptake and catecholamine secretion in primary dissociated cultures of adrenal medulla, *J. Neurochem.,* 39, 635, 1982.

95. **Uvnäs, B. and Aborg, C. H.,** Possible role of nerve impulse induced sodium ion flux in a proposed multivesicular fractional release of adrenaline and noradrenaline from the chromaffin cell, *Acta Physiol. Scand.,* 109, 363, 1980.

96. **Gusovsky, F., Hollingworth, E. B., and Daly, J. J.,** Regulation of phosphatidylinositol turnover in brain synaptoneurosomes: stimulatory effects of agents that enhance influx of sodium ions, *Proc. Natl. Acad. Sci. U.S.A.,* 83, 3003, 1986.

97. **Schneeweiss, F., Naquira, D., Rosenheck, K., and Schneider, A. S.,** Cholinergic stimulants and excess potassium ion increase the fluidity of plasma membranes isolated from adrenal chromaffin cells, *Biochim. Biophys. Acta,* 555, 460, 1979.

98. **Trifaró, J. M. and Lee, R. W. H.,** Actin and myosin in chromaffin cells: roles in cell function, in *Catecholamines: Basic and Clinical Frontiers,* Usdin, E., Kopin, J. J., and Barchas, J., Eds., Pergamon Press, New York, 1978, 358.

99. **Cooke, P. H. and Poisner, A. M.,** The role of cytoskeleton in adrenomedullary secretion, *Methods Achiev. Exp. Pathol.,* 9, 137, 1979.

100. **Schneider, A. S., Cline, H. T., Rosenheck, K., and Sonnenberg, M.,** Stimulus secretion coupling in isolated adrenal chromaffin cells: calcium channel activation and possible role of cytoskeletal elements, *J. Neurochem.,* 37, 567, 1981.

101. **Hall, P. F.,** The role of the cytoskeleton in endocrine function, in *Cellular Regulation of Secretion and Release,* Conn, P. M., Ed., Academic Press, New York, 1982, 195.

102. **Trifaró, J. M., Bader, M.-F., and Doucet, J.-P.,** Chromaffin cell cytoskeleton: its possible role in secretion, *Can. J. Biochem. Cell Biol.,* 63, 661, 1985.

103. **Gray, E. G.,** Neurotransmitter release mechanism and microtubules, *Proc. R. Soc. London,* 218, 253, 1983.

104. **Kondo, H., Wolosewik, J. J., and Pappas, G. D.,** The microtrabecular lattice of the adrenal medulla revealed by polyethylene glycol embedding and stereo electron microscopy, *J. Neurosci.,* 2, 57, 1982.

105. **Goldstein, I., Hoffstein, S., Gallin, J., and Weissmann, G.,** Mechanism of lysosomal enzyme release from human leukocytes: microtubule assembly and membrane fusion induced by a component of complement, *Proc. Natl. Acad. Sci. U.S.A.,* 70, 2916, 1973.

106. **Thoa, N. B., Wooten, G. F., Axelrod, J., and Koprin, I. J.,** Inhibition of release of dopamine-beta-hydroxylase and norepinephrine from sympathetic nerves by colchicine, vinblastine or cytochalasin B, *Proc. Natl. Acad. Sci. U.S.A.,* 69, 520, 1972.

107. **Trifaró, J. M., Collier, B., Lastowecka, A., and Stern, D.,** Inhibition by colchicine and vinblastine of acetylcholine-induced release from the adrenal gland: an anticholinergic action, but not an effect upon microtubules, *Mol. Pharmacol.,* 8, 284, 1972.

108. **McKay, D. B. and Schneider, A. S.,** Selective inhibition of cholinergic receptor-mediated $^{45}Ca^{2+}$ uptake and catecholamine secretion from adrenal chromaffin cells by taxol and vinblastine, *J. Pharmacol. Exp. Ther.,* 231, 102, 1984.

109. **Landry, Y., Amellal, M., and Ruckstuhl, M.,** Can calmodulin inhibitors be used to probe calmodulin effects?, *Biochem. Pharmacol.,* 30, 2031, 1981.

110. **Wada, A., Yanagihara, N., Izumi, F., Sakurai, S., and Kobayashi, H.,** Trifluoroperazine inhibits $^{45}Ca^{2+}$-uptake and catecholamine secretion and synthesis in adrenal medullary cells, *J. Neurochem.,* 40, 481, 1983.

111. **Powis, D.,** Clonidine inhibits catecholamine release from bovine adrenal medullary cells, but apparently not via alpha$_2$-adenoreceptors, *J. Physiol. (London),* 358, 87p, 1984.

112. **Lelkes, P. I. and Pollard, H. B.**, Inhibitors of metalloprotease activity block catecholamine release from bovine adrenal medullary cells by modulating calcium homeostasis, manuscript submitted.
113. **Whitaker, M. J. and Baker, P. F.**, Calcium-dependent exocytosis in an in vitro secretory granule plasma membrane preparation from sea urchin eggs and the effects of some inhibitors of cytoskeletal function, *Proc. R. Soc. London*, 218, 397, 1983.
114. **Morita, K., Levine, M., and Pollard, H. B.**, The possible role of the cytoskeletal network in exocytotic mechanism of catecholamine secretion from bovine adrenal medullary cells, personal communication.
115. **Lelkes, P. I., Friedmann, J. E., Rosenheck, K., and Oplatka, A.**, De-stabilization of actin filaments as a requirement for the secretion of catecholamines from permeabilized chromaffin cells, *FEBS Lett.*, in press.
116. **Friedman, J. E., Lelkes, P. I., Rosenheck, K., and Oplatka, A.**, The possible implication of membrane-associated actin in stimulus-secretion coupling in adrenal chromaffin cells, *Biophys. Biochem. Res. Commun.*, 96, 1717, 1980.
117. **Wehland, J., Weber, K., Gawlitta, W., and Stockem, W.**, Effects of the actin-binding protein DNAse I on cytoplasmic streaming and ultrastructure of *Amoeba proteus*, *Cell Tissue Res.*, 199, 353, 1979.
118. **Friedman, J. E.**, unpublished observation, 1982.
119. **Oplatka, A., Levy, R., and Friedman, J. E.**, The elementary unit of the actomyosin system, in *Biological Structures and Complex Flow*, Oplatka, A. and Balaban, M., Eds., Academic Press, New York and Balaban Publishers, Philadelphia, 1983, 207.
120. **Pollard, H. B., Creutz, C. E., Fowler, V. M., Scott, J. H., and Pazoles, C. J.**, Calcium-dependent regulation of chromaffin granule movement, membrane contact, and fusion during exocytosis, *Cold Spring Harbor Symp. Quant. Biol.*, 46, 819, 1983.
121. **Harish, O. E., Levy, R., Rosenheck, K., and Oplatka, A.**, Possible involvement of actin and myosin in Ca^{2+} transport trough the plasma membrane of chromaffin cells, *Biochem. Biophys. Res. Commun.*, 119, 652z, 1984.
122. **Friedman, J. E., Lelkes, P. I., Rosenheck, K., and Oplatka, A.**, Control of stimulus-secretion coupling in adrenal medullary chromaffin cells by microfilament specific macromolecules, *J. Biol. Chem.*, 261, 3745, 1986.
123. **Perrin, D. and Aunis, D.**, Reorganization of alpha-fodrin induced by stimulation in secretory cells, *Nature (London)*, 315, 589, 1985.
124. **Howard, J. H. and Meyer, W. H.**, Chemotactic peptide modulation of actin assembly and locomotion in neutrophils, *J. Cell Biol.*, 98, 1265, 1984.
125. **Wallace, P. J., Wresto, R. P., Packmann, C. H., and Lichtman, M. A.**, Chemotactic peptide-induced changes in neutrophil actin conformation, *J. Cell Biol.*, 99, 1060, 1984.
126. **Pfeiffer, J. R., Seagrave, J. C., Davis, B. H., Deanin, G. G., and Oliver, J. M.**, Membrane and cytoskeletal changes associated with IgE-mediated serotonin release from rat basophilic leukemia cells, *J. Cell Biol.*, 101, 2145, 1985.
127. **Lelkes, P. I.**, unpublished observation, 1985.
128. **Grinstein, S., Elder, B., and Furuya, W.**, Phorbol ester-induced changes of cytoplasmic pH in neutrophils: role of exocytosis in Na^+-H^+ exchange, *Am. J. Physiol.*, 248, C379, 1985.

Chapter 10

THE CHROMAFFIN CELL PLASMA MEMBRANE

Kurt Rosenheck

TABLE OF CONTENTS

I. INTRODUCTION

In the search for appropriate experimental systems for studying the mechanism of secretion of adrenal medullary catecholamines, the chromaffin cell plasma membrane (PM) is a relative latecomer. Work describing the preparation of PM began to appear about 10 years ago. In contrast to the ease in obtaining highly purified fractions of chromaffin granules (see Chapter 3), the methods that are used for the fractionation of PM are still somewhat lacking in efficacy. On the other hand, PM is an indispensable and fascinating object for the understanding of the stimulus-secretion response, being the site which comprises both the early and the ultimate links in the chain of molecular processes constituting the latter. While some work with PM has dealt in recent years with the very last facet, i.e., exocytotic fusion, the stimulus-related trigger events such as ionic channel functions and lipolytic transformations have hardly been touched. An outlook on these possibilities will be given in Section IV. I begin with an outline of the methods for the preparation of PM, their biochemical and ultrastructural characterization, and their use in looking at interactions with cytoplasmic proteins or chromaffin granules, modeling exocytosis.

II. PLASMA MEMBRANE PREPARATION AND ANALYSIS

Subsequent to the removal of cell debris and subcellular organelles by preliminary centrifugation steps, practically all the methods developed for PM fractionation use either continuous or discontinuous sucrose gradients for the final purification step.[1-8] In the Wilson and Kirshner procedure,[2,3] the final centrifugation is done on a discontinuous sucrose-renographin gradient.

A. Purity

The plasma membrane marker enzyme acetylcholinesterase (EC 3.1.1.7) has been mainly used to estimate enrichment, but other markers have also been determined. Table 1 lists the procedures and enrichment factors. The latter is fairly constant at $\sim$12 with a maximum variation of $\sim$20%. 5′-Nucleotidase (EC 3.1.3.5) has frequently been measured in PM,[1,2,4,6] but enrichment factors calculated vary considerably (between 3 and 25), and its reliability as a PM marker has been questioned.[2] The Ca^{2+}-dependent ATPase (EC 3.6.1.3) is more consistent, with enrichment factors in the region of 10.[1,2,8,9] Adenylate cyclase is enriched about four- to fivefold with respect to the homogenate,[2,4,5] and a similar value was found for guanylate cyclase.[5]

The phospholipid composition of purified PM appears in Table 2. Except for the PS values, which differ significantly, the analyses are in good agreement. The PM has a relatively high SM content and lower cholesterol/phospholipid ratio, as compared to chromaffin granule membranes.

B. Membrane Proteins

The nicotinic acetylcholine receptor of bovine adrenal medullary PM was studied by Wilson and Kirshner,[3] by determining specific binding of [^{125}I]α-bungarotoxin to the PM fractions, as well as by perfusion experiments of isolated glands. The nicotinic nature of the receptor was shown by the inhibitory action of the nicotinic antagonists decamethonium and hexamethonium on both secretory response and specific α-bungarotoxin binding. Oddly enough, the classic muscarinic antagonist, atropine, showed a similar effect, when applied at concentrations in the millimolar ranges, i.e., higher than those used to block muscarinic responses. Scatchard plots of α-bungarotoxin binding to PM gave a dissociation constant of 1.6 nM and a concentration of 190 fmol of binding sites per milligram membrane protein.[3] Muscarinic receptors have been found in a nonfractionated membrane preparation from

Table 1
**SOME REPRESENTATIVE PROCEDURES FOR THE PREPARATION OF
PLASMA MEMBRANE FROM ADRENAL MEDULLARY CHROMAFFIN CELLS**

Final separation steps of plasma membrane fraction	Enrichment factor (AChE activity relative to homogenate)	Ref.
Zonal centrifugation on continuous sucrose gradient	14	1
Centrifugation on 1.4 M sucrose, followed by discontinuous gradient of 7, 10, and 12% Renografin in 0.3 M sucrose	13	2, 3
Centrifugation on discontinuous sucrose gradient (d = 1.08, 1.14, 1.21, and 1.29 g/mℓ), followed by linear sucrose gradient (d = 1.08 to 1.14 g/mℓ)	Not determined (see text for other marker enzymes)	4
Centrifugation on three-step Percoll gradient (20, 40, and 60%), followed by discontinuous sucrose gradient (20, 32, and 36%)	10	6
Centrifugation on 1.4 M sucrose, followed by centrifugation on 1.0 M sucrose	11	8

Table 2
**LIPID COMPOSITION OF ADRENAL
MEDULLARY PLASMA MEMBRANE
FRACTIONS (%)**

	Ref. 1	Ref. 2	Ref. 8
Phosphatidylcholine (PC)	38.1	34.4	35.4
Lysophosphatidylcholine (LPC)	4.3	1.4	4.2
Phosphatidylethanolamine (PE)	28.5	35.7	27.7
Phosphatidylserine (PS)	4.9	12.4	9.4
Phosphatidylinositol (PI)	1.1	3.0	3.5
Sphingomyelin (SM)	17.1	15.4	14.5
Phosphatidic acid (PA)	1.0	<0.2	1.7
Cholesterol (mg/mg protein)		0.06	0.073

homogenized medullae.[10] K_D, in this case was 0.07 nM and B_{max} was ~3 fmol/mg membrane protein. This subject is treated in more detail in Chapter 11.

Tubulin was detected in purified PM by SDS gel electrophoresis, coupled with fingerprint analysis,[4] and actin was found by the same method[4] and by the DNAse assay.[11] The PM-actin linkage is stable to treatment with Mg^{2+}-ATP.[4] Actin was also present in a tenfold-enriched (by acetylcholinesterase assay) PM fraction prepared from bovine chromaffin cells by isolation on polycationic beads,[12] according to the method of Jacobson and Branton.[13] Actin is almost certainly associated with the PM, probably via interactions with spectrin-like protein, present on the cytoplasmic face.[14] Another plasma membrane protein of an apparent molecular weight of about 51,000 on SDS-β mercaptoethanol gels was isolated by affinity chromatography on a glutaraldehyde-fixed chromaffin granule column.[6] The authors suggest its possible functioning as a plasma membrane-located anchoring molecule for the chromaffin granules during the exocytotic process. Creatine kinase (EC 2.7.3.2) has been found in the adrenal medullary PM[11] and used to assess the sidedness of the PM vesicles, as a convenient marker for the cytoplasmic surface of the membrane. It is purely BB (brain) type, similar to that identified in the acetylcholine receptor-rich membranes from the *Torpedo* electric organ,[15] as well as in erythrocyte ghosts.[16] Another protein associated with the PM is a metalloendoprotease,[17,18] which was found to be located on the cytoplasmic membrane

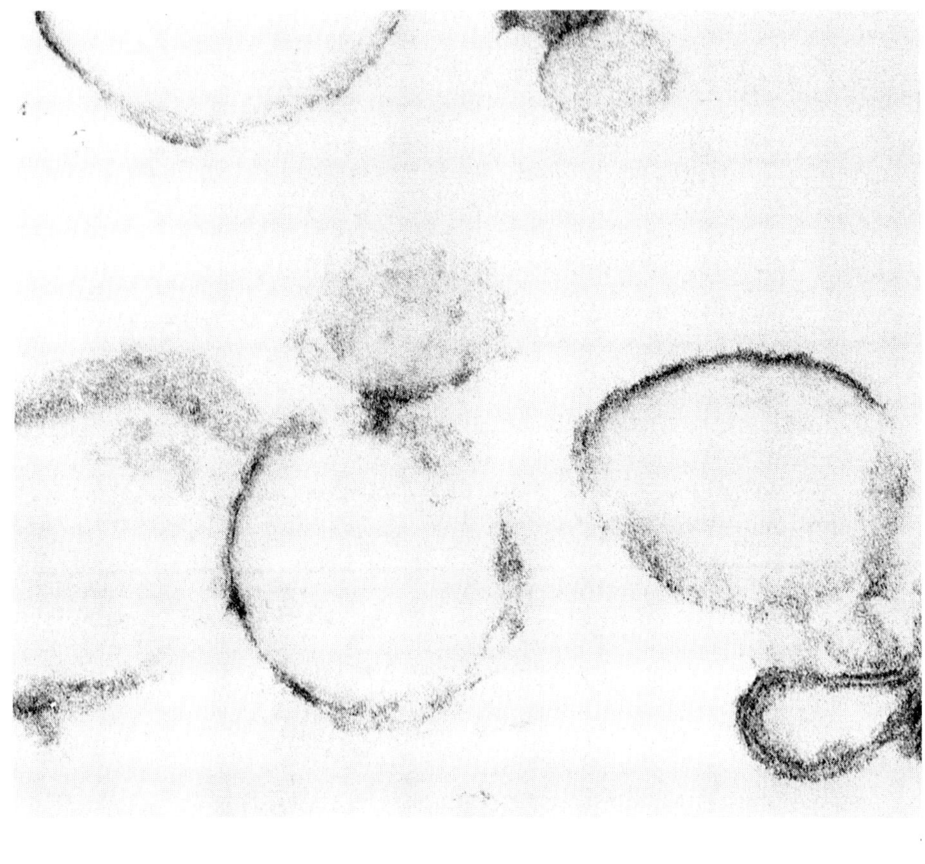

A

FIGURE 1.   Impermeability of plasma membrane vesicles to Ruthenium Red (M_r = 800). The membranes were incubated with the stain[48] in the absence (A) or presence (B) of 0.01% Saponin, prior to fixation in 1.5% glutaraldehyde, followed by 2% OsO_4, dehydration in acetone, embedding in Spurr's resin, and staining of ultrathin sections with 2% uranyl acetate and alkaline lead citrate. (Magnification × 160,000.)

surface.[19] Inhibitors of this enzyme were found to interfere with a series of cellular functions, including Ca^{2+} flux,[17] protein phosphorylation,[17] and catecholamine secretion.[17,18] Cytochrome *b*-561, which is a major constituent of the chromaffin granule membrane (see Chapter 3), has also been found in purified PM fractions by difference spectrophotometry.[20,21] Its specific activity is about 15 to 25% of that found in the granule membranes. The functional role of this enzyme in plasma membranes is so far not known. A high-affinity binding site for prostaglandin E_2 has been found in purified bovine PM.[9] Its concentration was 2.23 pmol/mg protein, higher than that of any of the other subcellular fractions of the adrenal medulla. The dissociation constant was 1.9 nM. This site possibly mediates the inhibitory effect of prostaglandin E_2 on catecholamine release from the adrenal medulla.[22]

C. Ultrastructure

Electron microscopy of PM fractions has shown them to be composed of smooth vesicles, varying in diameter from ~0.1 to ~0.5 μm.[1,2,4,6,8] The majority of these vesicles are closed and impermeable to cytochemical markers the size of Ruthenium Red (M_r = 800).[23] Figure 1 shows electron micrographs of a PM fraction, prepared according to the method of Wilson and Kirshner[2] and stained with Ruthenium Red, with or without prior permeabilization by

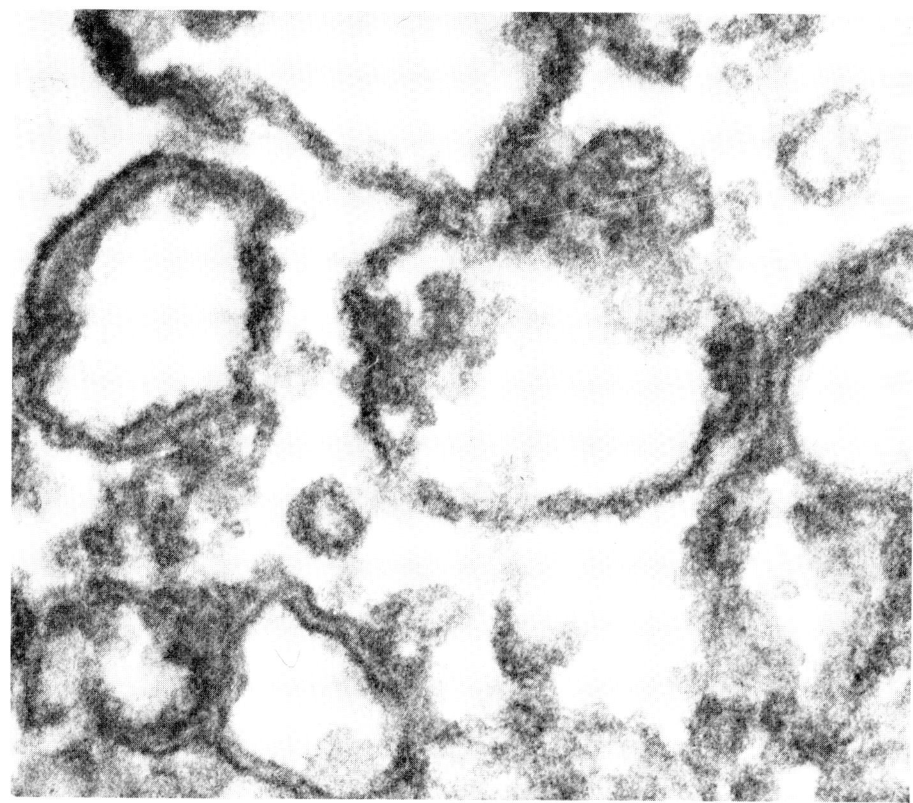

FIGURE 1B.

0.01% Saponin. In the absence of detergent there is practically no stain to be found on the inner surface of the vesicles. Only a very minor fraction is permeable to horseradish peroxidase (M_r = 40 kdaltons), and none of the vesicles is permeable to markers larger than this.[23] The sidedness of the membrane vesicles has been determined using cytochemical markers for the outer and inner surface.[23] Figure 2 shows a freeze-etching micrograph of PM vesicles treated with heavy meromyosin in order to enhance the visualization of actin filaments associated with the membrane surfaces. Most of the vesicles present smooth outer and inner surfaces, indicating that they are closed and oriented right-side out. One only is seen to be studded on the outside with heavy meromyosin and thus probably is turned inside out.[23] Use of this and other markers gave results consistent with the conclusion that most of the vesicles are oriented right-side out. This conclusion is a confirmation of an earlier study of membrane sidedness,[11] in which the accessibility of two markers each, for the external and the cytoplasmic membrane surface, respectively, was measured. Table 3 lists the main results of this study, which show that about 85% of the vesicles are sealed, and that about 95% of these are turned right-side out.

III. FUNCTIONAL STUDIES

A. Stimulus-Related
1. Phosphoinositide Turnover
Azila and Hawthorne[8] have looked at the changes in phosphatidylinositol and phosphatidate in subcellular fractions prepared from the medulla of perfused bovine adrenal glands, in

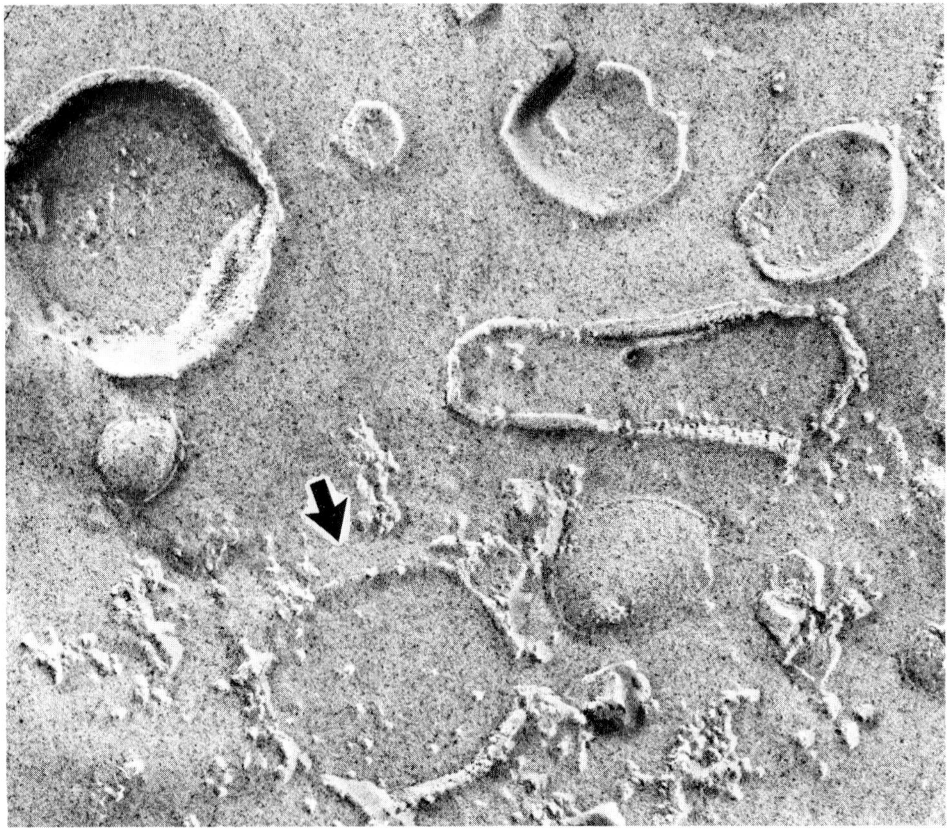

FIGURE 2.   Visualization of membrane-associated actin by decoration with heavy meromyosin. One vesicle is decorated on the outside (arrow). The sample was incubated with heavy meromyosin in 40 mM NaCl, 5 mM phosphate buffer, pH 6.5 for 3 hr and pelleted, prior to preparation for freeze fracturing by cryofixation.[49,50] (Magnification × 100,000.)

which secretion was stimulated by either carbachol or excess KCl, after previous labeling with [32]P. The specific radioactivities of these two components decreased by about 50% after carbachol stimulation, while those of other phospholipids were unaffected. However, quantitatively similar changes were measured in all subcellular fractions (granule, mitochondrial, microsomal, and plasma membranes) and thus do not necessarily reflect a lipid response occurring exclusively in the PM. Furthermore, muscarinic stimulation did not lead to significant phospholipid changes. It is of interest that KCl-induced secretion was not accompanied by phospholipid changes, while carbachol in the absence of external Ca^{2+} did bring about changes. Thus, the phospholipid response seems to be clearly related to cholinergic receptor activation.[8] Comprehensive studies dealing with phospholipid metabolism are being carried out with intact and leaky cells and are described in Chapter 11. Very recently, we have started work on arachidonic acid production in PM with the purpose of studying the activation of phospholipase A_2 in these membranes. In the course of this study a membrane-associated PI-specific phospholipase C was detected, as well as a diglyceride lipase, which might be responsible for arachidonic acid formation directly from the diacylglycerol produced in the PI cycle.[24]

2. Membrane Fluidity

Chromaffin cell PM suspensions, labeled with the hydrophobic fluorescent probe 1,6-

Table 3
ASSAY OF THE SIDEDNESS OF
PLASMA MEMBRANE VESICLES,
PREPARED ACCORDING TO WILSON
AND KIRSHNER[2,7]

Marker proteins	Accessibility (%)
External surface	
α-Bungarotoxin binding sites	95 ± 4
Acetylcholinesterase activity	98 ± 2
Cytoplasmic surface	
Actin	20 ± 2
Creatine kinase activity	22 ± 6

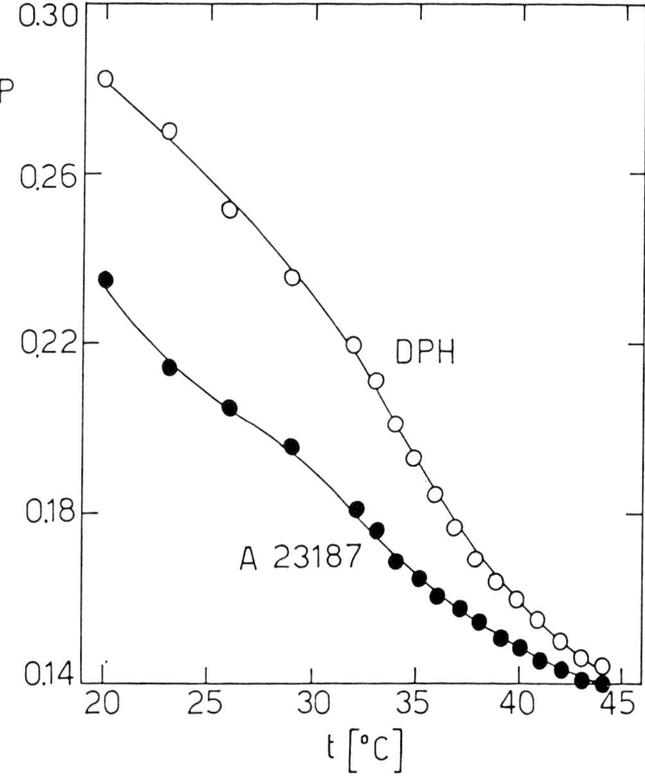

FIGURE 3.   Fluorescence polarization, P, as function of temperature, t, of plasma membrane vesicles, measured with an "Elscint Microviscosimeter"[7] using either DPH[51] or A23187[52,53] as fluorescence probes. The "phase transition" appears at about the same temperature (30°C) in both these cases.

diphenyl-1,3,5-hexatriene, undergo remarkable changes in fluorescence polarization as a function of both temperature and the presence or absence of nicotinic agonist.[7] Figures 3 and 4 demonstrate these changes. The progressive decrease in P values as temperature is raised, or agonist added, signals fluidization of the lipid bilayer.[25] The main results of this study were the presence of a break at ~30°C in the P-vs.-t profiles, indicating a phase transformation from gel-like to more liquid-crystalline lipid regions within the membrane,

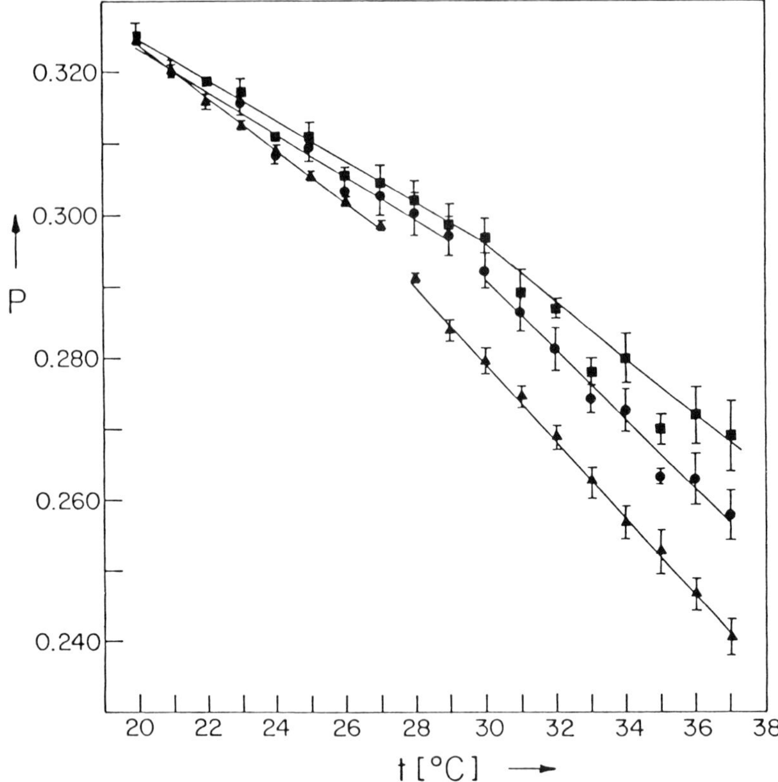

FIGURE 4.   Effect of cholinergic agonists, in the presence of 2 mM Ca^{2+}, on the temperature dependence of the degree of fluorescence polarization of DPH in plasma membrane vesicles. Squares: no agonist; circles: 0.5 mM acetylcholine; triangles: 50 μM nicotine. Measurements performed as in Figure 3. The variation of P in the temperature ranges below and above the ''phase transition'' is approximated by straight lines.

and increases in flow activation energy by ~1 and ~3 kcal/mol in the presence of acetylcholine and nicotine, respectively. The cholinergic inhibitor hexamethonium abolished these shifts in flow activation energy. Furthermore, excess K$^+$ produced similar changes in fluidity. Two complementary interpretations were proposed, one dealing with the putative reshuffling of receptor-associated lipids under the influence of agonist binding to the receptor, the other with effects of the ionic redistribution and altered membrane potential on the organization of lipids. These latter effects could arise from either excess K$^+$ or agonist-induced opening of the ionic channel, linked to the receptor. This possibility required the demonstration that the vesicles are able to maintain ionic gradients. Support for this assumption comes from a study described below in which the membrane potential of PM subjected to varying K$^+$ gradients was measured with the fluorescent probe DiS-C$_3$-(5). Another study of plasma membrane fluidity, using two nitroxide spin labels, could confirm the existence of a break situated at ~26°C, in the temperature-dependence profiles of the rotational correlation times, but failed to find any effect of acetylcholine on these quantities.[26] The discrepancy between the spin label and fluorescence polarization measurements may be due to the possibility that these probes sample different regions of the lipid bilayer. The changes in agonist-induced lipid fluidity, monitored by the fluorescence probe, may arise in the close vicinity of the cholinergic receptor. It has been concluded elsewhere that 1,6-diphenyl-1,3,5-hexatriene partitions into the interface between lipid domains.[27] It thus appears that the nitroxide spin labels are not suitable for the detection of agonist-induced modulations of the lipid distribution

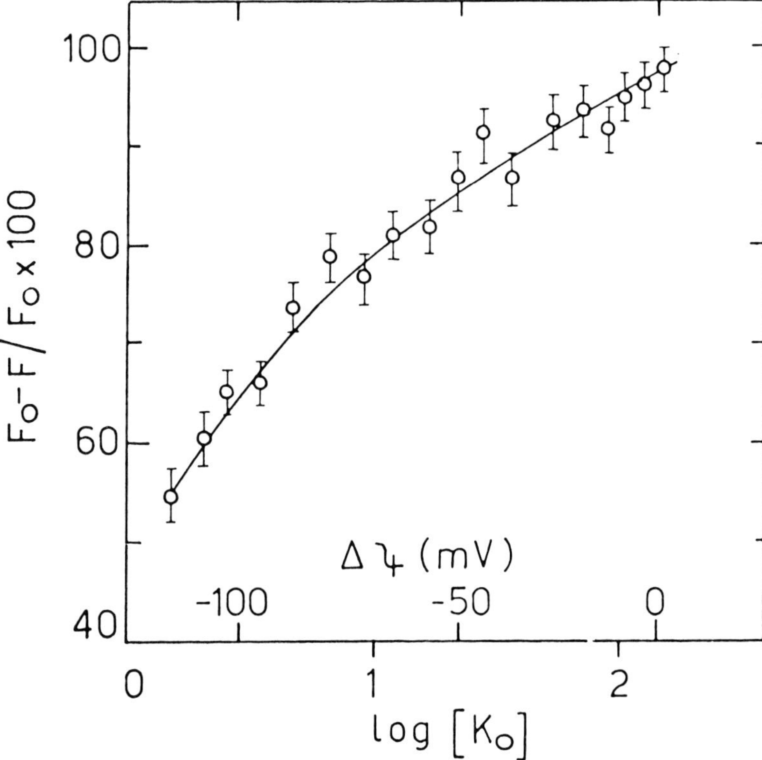

FIGURE 5.   Relative fluorescence intensity changes of the transmembrane potential probe 3,3′-dipropylthiadicarbocyanine, DiS-C_3-(5),[54] in plasma membrane vesicles[11] as function of membrane potentials imposed by varying the external K^+ concentration. The internal medium consisted of 150 mM KCl, which was loaded into the vesicles using the French Press method.[55] $\Delta\Psi$ Was calculated by the Nernst equation (courtesy of P. I. Lelkes).

between gel-like and liquid-crystalline domains in the PM. Furthermore, these changes may reflect stimulus-related events, rather than processes connected with exocytotic membrane fusion. On the other hand, the effects of K^+, which are additive to those of the cholinergic agonist,[7] most probably reflect a larger-scale reorganization of membrane lipids.

3. Membrane Potential

Figure 5 shows the response of the DiS-C_3-(5) fluorescence to a varying K^+-induced Nernst potential,[28] indicating that the PM vesicles are sealed and can maintain an ionic gradient. Addition of acetylcholine partially depolarizes the vesicle membrane, presumably, as a consequence of the opening of the receptor-associated ion channel.[11] Similar effects can be observed in the presence of ionophores, such as Gramicidin-D.[11] This property of the PM is extremely interesting, since it provides us with the possibility to study the ionic membrane channels involved in stimulation by tracer flux experiments.

B. Exocytosis Related

1. Synexin Binding

The calcium-binding protein synexin (see Chapter 8) is being thought to act as a mediator in exocytosis,[29] by its ability to enhance chromaffin granule aggregation.[30] Its interaction with PM was investigated by exposing the cytoplasmic surface of lysed chromaffin cells to the medium, using the polycationic bead technique.[13,31] Synexin was found to bind in a calcium-dependent manner, the half-maximal binding occurring at a calcium concentration of about 200 μM. The calcium titration curve was similar to those for granule aggregation and synexin self-association. The significance of these findings has to be examined further, since the calcium activation curve for binding lies at higher concentrations than the internal free calcium levels that trigger exocytotic secretion in the intact cell (see Chapter 2).

2. PM-Induced Granule Lysis

The earliest account of catecholamine release from granules incubated with plasma membranes, isolated from the adrenal medulla of rabbits, is preliminary experiments performed in the framework of a more general study of insulin release from β-granules.[32] Release was found to require 2 μM Ca^{2+}. Somewhat later, release from bovine chromaffin granules was found to occur in the presence of a releasing factor partially associated with the microsomal fraction of the chromaffin cell, that could be solubilized by treatment with cholic acid.[33] It is not clear whether this protein was a constituent of the plasma membrane. The Ca^{2+} requirement of this release was not studied. In a series of investigations over recent years,[34-39] Konings and De Potter established the calcium requirement for the release from bovine granules incubated with PM, isolated by the procedure of Meyer and Burger.[6] Maximal catecholamine release was reached at 10 μM Ca^{2+}, the half-maximal response being at about 1 μM.[35] The amount of catecholamine released constituted about 30% of the total catecholamines present in the assay. Other granule constituents (e.g., ATP, DBH) were released simultaneously,[36] indicating all-or-none lysis of the granules. This conclusion was confirmed by use of differential centrifugation of the interaction products on Percoll gradients.[37] A requirement for sialic acid was shown by the inhibitory action of Ruthenium Red, or pretreatment with neuraminidase, on release,[38] as well as concomitant phosphorylation of two PM proteins, having apparent molecular weights of about 20 and 60 kdaltons.[39] The relevance of these findings within the framework of the search for suitable model systems for the exocytotic event remains to be established, particularly in view of the fact that the sidedness of the PM, prepared according to the procedure of Meyer and Burger,[6] has not been determined in these studies.

3. Chromaffin Granule-PM Interactions

Assays were carried out using differential centrifugation on either Percoll gradients[40] or a discontinuous sucrose gradient,[11,41] and monitoring the sedimentation of markers (e.g., acetylcholinesterase) characteristic for one or the other of the components. After centrifugation in Percoll, complexes were detected at both low (0.1 μM) and high (10 to 100 μM) Ca^{2+} levels, but only those at high Ca^{2+} concentration were dissociable by 1 mM Mg-ATP.[40] This apparent independence of the interaction on the Ca^{2+} level was also found in a fluorimetric study, using as a monitor the quenching of the fluorescence of 6-carboxyfluorescein contained in the PM vesicles by the catecholamines of intact chromaffin granules.[42] The interaction, interpreted as fusion between the two components, was triggered by 0.5 mM acetylcholine, irrespective of whether 2.2 mM Ca^{2+} was present or absent. However, since this study was not performed in the presence of calcium buffers, this apparent discrepancy with the physiologic calcium requirement might have been due to random Ca^{2+} leaks from the granules during the experiment, sufficient to elevate local Ca^{2+} levels to the micromolar or higher range. A further question was raised by the later finding that the large majority of PM vesicles has right-side-out orientation,[11] while fusion with granules would be expected to occur with the cytoplasmic face of the PM. Possibly the small fusion yields of ~5% observed in the fluorescence work,[42] as well as in sucrose gradient centrifugation experiments on acetylcholinesterase transfer,[11,41] are a consequence of this situation. The discrepancy between the very small effects in the aforementioned studies and the much larger ones in the catecholamine release assays[34,35] is unresolved at present.[43]

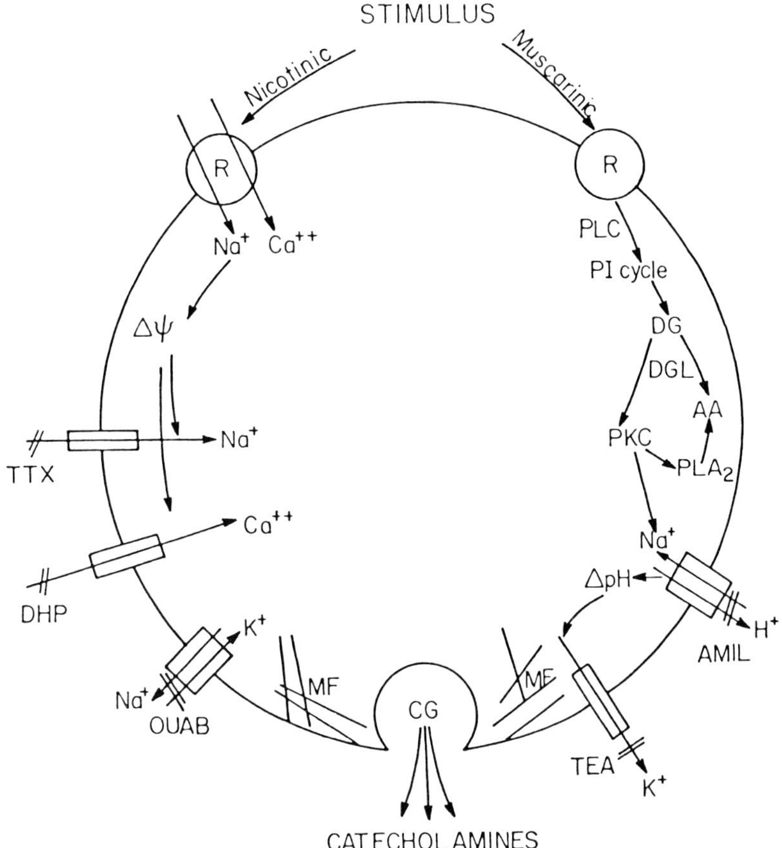

FIGURE 6.   Chart of part of the events occurring within, across, or in the close neighborhood of the plasma membrane in adrenal medullary cells, some well established, some putative. Many of the symbols are self-explanatory. Arrows usually signify "activation". R: receptor. Inhibitors: TTX, tetrodotoxin; DHP, dihydropyridines; OUAB, ouabaine; TEA, tetraethylammonium; AMIL, amiloride. Enzymes: PLC, phospholipase C; DGL, diacylglycerol lipase; PKC, protein kinase C; PLA$_2$, phospholipase A$_2$. Lipids: PI, phosphatidylinositol; DG, diacyglycerol; AA, arachidonic acid. MF: membrane-associated microfilaments. CG: chromaffin granule.

IV. OUTLOOK

In addition to serving as a model in reconstitution experiments of exocytotic secretion, plasma membrane vesicles hold great promise as an object for the study of stimulus-related events. Figure 6 presents some of the components of the membrane or in its close neighborhood that are worthy of examination. While a part of these, representing the nicotinic pathway, has been studied in some detail in the intact cell and by electrophysiological techniques, the others, including muscarinic mechanisms and possible cross relations between the two sides, are just beginning to be charted. Further pathways to be considered are Na$^+$-Ca^{2+} exchange[44-46] and Ca^{2+} extrusion pumps.[47]

Using vesicles, experimental approaches may be found to still unresolved problems such as: (1) lypolytic pathways in stimulus-secretion; (2) mutual regulatory roles of membrane-associated cytoskeletal elements as well as cytoplasmic proteins, and membrane lipids; (3) lipid requirements of exocytosis. Data on questions such as these could also be relevant to the understanding of channel and receptor desensitization and the termination of the secretory response (see Chapters 14 and 15).

For these and other experiments, methods will have to be worked out for the convenient isolation of purified plasma membrane fractions from the relatively small cell samples that can be used in stimulation experiments. The fact that the plasma membrane vesiculates spontaneously right-side out in an almost quantitative way will be of great help in the study of the stimulus-related ion fluxes, for example. When the object is exocytotic fusion, methods for an effective reversal of the membrane orientation will have to be found.

ACKNOWLEDGMENT

The author's work has been supported in part by the Stiftung Volkswagenwerk, as well as by fellowships conferred to him and to some of his colleagues by the D.A.A.D. and the Minerva Stiftung, Heidelberg, West Germany.

REFERENCES

1. **Nijjar, M. S. and Hawthorne, J. N.,** A plasma membrane fraction from bovine adrenal medulla: preparation, marker enzyme studies and phospholipid composition, *Biochim. Biophys. Acta,* 367, 190, 1974.
2. **Wilson, S. P. and Kirshner, N.,** Isolation and characterization of plasma membranes from the adrenal medulla, *J. Neurochem.,* 27, 1289, 1976.
3. **Wilson, S. P. and Kirshner, N.,** The acetylcholine receptor of the adrenal medulla, *J. Neurochem.,* 28, 687, 1977.
4. **Zinder, O., Hoffman, P. G., Bonner, W. M., and Pollard, H. B.,** Comparison of chemical properties of purified plasma membranes and secretory vesicle membranes from the bovine adrenal medulla, *Cell Tissue Res.,* 188, 153, 1978.
5. **Aunis, D., Pescheloche, M., and Zwiller, J.,** Guanylate cyclase from bovine adrenal medulla: subcellular distribution and studies on the effect of lysolecithin on enzyme activity, *Neuroscience,* 3, 83, 1978.
6. **Meyer, D. I. and Burger, M. M.,** Isolation of a protein from the plasma membrane of adrenal medulla which binds to secretory vesicles, *J. Biol. Chem.,* 254, 9854, 1979.
7. **Schneeweiss, F., Naquira, D., Rosenheck, K., and Schneider, A. S.,** Cholinergic stimulants and excess potassium ion increase the fluidity of plasma membranes isolated from adrenal chromaffin cells, *Biochim. Biophys. Acta,* 555, 460, 1979.
8. **Azila, N. and Hawthorne, J. N.,** Subcellular localization of phospholipid changes in response to muscarinic stimulation of perfused bovine adrenal medulla, *Biochem. J.,* 204, 291, 1982.
9. **Karaplis, A. C. and Powell, W. S.,** Subcellular localization of prostaglandin E_2 binding sites in bovine adrenal medulla, *Biochim. Biophys. Acta,* 801, 189, 1984.
10. **Kayaalp, S. O. and Neff, N. H.,** Cholinergic muscarinic receptors of bovine adrenal medulla, *Neuropharmacology,* 18, 909, 1979.
11. **Lelkes, P. I., Naquira, D., Friedman, J. E., Rosenheck, K., and Schneider, A. S.,** Plasma membrane vesicles from bovine adrenal chromaffin cells: characterization and fusion with chromaffin granules, in *Advances in the Biosciences,* Vol. 36, Izumi, F. et al., Eds., Pergamon Press, Oxford, 1982, 143.
12. **Van der Meulen, J. A., Emerson, D. M., and Grinstein, S.,** Isolation of chromaffin cell plasma membranes on polycationic beads, *Biochim. Biophys. Acta,* 643, 601, 1981.
13. **Jacobson, B. S. and Branton, D.,** Plasma membrane rapid isolation and exposure of the cytoplasmic surface by positively charged beads, *Science,* 195, 302, 1977.
14. **Aunis, D. and Perrin, D.,** Chromaffin granule membrane-F-actin interactions and spectrin-like protein of subcellular organelles: a possible relationship, *J. Neurochem.,* 42, 1558, 1984.
15. **Barrantes, F. J., Mieskes, G., and Wallimann, T.,** Creatine kinase activity in the *Torpedo* electrocyte and in the nonreceptor, peripheral proteins from acetylcholine receptor-rich membranes, *Proc. Natl. Acad. Sci. U.S.A.,* 80, 5440, 1983.
16. **Mawatari, S. and Shinnoh, N.,** Occurrence of creatine kinase activity in human erythrocyte membranes, *Biochim. Biophys. Acta,* 643, 669, 1981.
17. **Lelkes, P. I., Brocklehurst, K. M., Morita, K., and Pollard, H. B.,** Metalloendoproteases in bovine adrenal chromaffin cells: their localization and involvement in stimulus-secretion coupling, presented at Int. Symp. on Molecular Biology of Peripheral Catecholamine Storing Tissues, Colmar, France, August 5 to 9, 1984.

18. **Mundy, D. I. and Strittmatter, W. J.**, Requirement for metalloendoprotease in exocytosis: evidence in mast cells and adrenal chromaffin cells, *Cell*, 40, 645, 1985.
19. **Lelkes, P. I.**, personal communication, 1985.
20. **Malviya, A. N., Rendon, A., and Aunis, D.**, Interaction of antimycin with cytochrome b-561. A study in secretory granules and in plasma membrane isolated from chromaffin cells of bovine adrenal medulla, *FEBS Lett.*, 160, 153, 1983.
21. **Rosenheck, K. and Plattner, H.**, unpublished data, 1984.
22. **Gutman, Y. and Boonyaviroj, P.**, Mechanism of PGE inhibition of catecholamine release from adrenal medulla, *Eur. J. Pharmacol.*, 55, 129, 1979.
23. **Rosenheck, K. and Plattner, H.**, *Biochim. Biophys. Acta*, 856, 373, 1986.
24. **Zahler, P., Reist, M., Pilarska, M., and Rosenheck, K.**, *Biochim. Biophys. Acta*, 877, 372, 1986.
25. **Shinitzky, M. and Yuli, I.**, Lipid fluidity at the submacroscopic level: determination by fluorescence polarization, *Chem. Phys. Lipids*, 30, 261, 1982.
26. **Rimle, D., Morse, P. D., II, and Njus, D.**, A spin-label study of plasma membranes of adrenal chromaffin cells, *Biochim. Biophys. Acta*, 728, 92, 1983.
27. **Klausner, R. D., Kleinfeld, A. M., Hoover, R. L., and Karnovsky, M. J.**, Lipid domains in membranes, *J. Biol. Chem.*, 255, 1286, 1980.
28. **Lelkes, P. I. and Friedman, J.**, personal communication, 1982.
29. **Pollard, H. B., Pazoles, C. J., Creutz, C. E., and Zinder, O.**, The chromaffin granule and possible mechanisms of exocytosis, *Int. Rev. Cytol.*, 58, 159, 1979.
30. **Creutz, C. E., Pazoles, C. J., and Pollard, H. B.**, Identification and purification of an adrenal medullary protein (synexin) that causes calcium-dependent aggregation of isolated chromaffin granules, *J. Biol. Chem.*, 253, 2858, 1978.
31. **Scott, J. H., Creutz, C. E., Pollard, H. B., and Ornberg, R.**, Synexin binds in a calcium-dependent fashion to oriented chromaffin cell plasma membranes, *FEBS Lett.*, 180, 17, 1985.
32. **Davis, B. and Lazarus, N. R.**, An *in vitro* system for studying insulin release caused by secretory granules-plasma membrane interaction: definition of the system, *J. Physiol. (London)*, 256, 709, 1976.
33. **Izumi, F., Kashimoto, T., Miyashita, T., and Wada, H.**, Involvement of membrane associated protein in ADP-induced lysis of chromaffin granules, *FEBS Lett.*, 78, 177, 1977.
34. **Konings, F. and De Potter, W.**, Calcium-dependent *in vitro* interaction between bovine adrenal medullary cell membranes and chromaffin granules as a model for exocytosis, *FEBS Lett.*, 126, 103, 1981.
35. **Konings, F. and De Potter, W.**, In vitro interaction between bovine adrenal medullary cell membranes and chromaffin granules: specific control by Ca^{2+}, *Naunyn-Schmied. Arch. Pharmacol.*, 317, 97, 1981.
36. **Konings, F. and De Potter, W.**, The chromaffin granule-plasma membrane interaction as a model for exocytosis: quantitative release of the soluble granule content, *Biochem. Biophys. Res. Commun.*, 104, 254, 1982.
37. **Konings, F., Majchrowicz, B., and De Potter, W.**, Release of chromaffin granular content on interaction with plasma membranes, *Am. J. Physiol.*, 254, C309, 1983.
38. **Konings, F. and De Potter, W.**, A role for sialic acid containing substrates in the exocytosis-like *in vitro* interaction between adrenal medullary plasma membranes and chromaffin granules, *Biochem. Biophys. Res. Commun.*, 106, 1191, 1982.
39. **Konings, F. and De Potter, W.**, Protein phosphorylation and the exocytosis-like interaction between adrenal medullary plasma membranes and chromaffin granules, *Biochem. Biophys. Res. Commun.*, 110, 55, 1983.
40. **Grafenstein, H. and Neumann, E.**, Interaction of chromaffin granules with plasma membranes mediated by Ca^{2+} and Mg^{2+}-ATP using self-generating gradients of Percoll, *FEBS Lett.*, 123, 238, 1981.
41. **Naquira, D., Lelkes, P. I., Lavie, E., and Rosenheck, K.**, *In vitro* fusion between isolated chromaffin cell membranes and granules: a model for exocytotic catecholamine release, in 13th FEBS Meet. Abstracts, Jerusalem, August 24 to 29, 1980, 247.
42. **Lelkes, P. I., Lavie, E., Naquira, D., Schneeweiss, F., Schneider, A. S., and Rosenheck, K.**, Acetylcholine-induced *in vitro* fusion between cell membrane vesicles and chromaffin granules from the bovine adrenal medulla, *FEBS Lett.*, 115, 129, 1980.
43. **Bental, M.**, The Interaction of Chromaffin Granules and Chromaffin Granule Ghosts with Natural and Model Membranes, M.Sc. thesis, The Feinberg Graduate School, The Weizmann Institute of Science, Rehovot, Israel 1983.
44. **Banks, P., Biggins, R., Bishop, R., Christian, B., and Currie, N.**, Sodium ions and the secretion of catecholamines, *J. Physiol. (London)*, 200, 797, 1969.
45. **Rink, T. J.**, The influence of sodium on calcium movements and catecholamine release in thin slices of bovine adrenal medulla, *J. Physiol. (London)*, 266, 297, 1977.
46. **Sorimachi, M., Nishimura, S., and Yamagami, K.**, Possible occurrence of Na^+-dependent Ca^{2+} influx mechanism in isolated bovine chromaffin cells, *Brain Res.*, 208, 442, 1981.

47. **Leslie, S. W. and Borowitz, J. L.,** Evidence for a plasma membrane calcium pump in bovine adrenal medulla but not adrenal cortex, *Biochim. Biophys. Acta,* 394, 227, 1975.
48. **Luft, J. H.,** Ruthenium Red and Violet. Chemistry, verification, methods of use for electron microscopy and mechanism of action, *Anat. Rec.,* 171, 347, 1971.
49. **Pscheid, P., Schudt, C., and Plattner, H.,** Cryofixation of monolayer cell cultures for freeze-fracturing without chemical pre-treatments, *J. Microsc. (Oxford),* 121, 149, 1981.
50. **Knoll, G., Oebel, G., and Plattner, H.,** A simple sandwich-cryogen-jet procedure with high cooling rates for cryofixation of biological materials in the native state, *Protoplasma,* 111, 161, 1982.
51. **Shinitzky, M. and Inbar, M.,** Microviscosity parameters and protein mobility in biological membranes, *Biochim. Biophys. Acta,* 433, 133, 1976.
52. **Case, G. D., Vanderkooi, J. M., and Scarpa, A.,** Physical properties of biological membranes determined by the fluorescence of the calcium ionophore A23187, *Arch. Biochem. Biophys.,* 162, 174, 1974.
53. **Weidekamm, E., Schudt, C., and Brdiczka, D.,** Physical properties of muscle cell membranes during fusion. A fluorescence polarization study with the ionophore A23187, *Biochim. Biophys. Acta,* 443, 169, 1976.
54. **Waggoner, A. S.,** Dye indicators of membrane potential, *Ann. Rev. Biophys. Bioenerg.,* 8, 47, 1979.
55. **Barenholzt, Y., Amselem, S., and Lichtenberg, D.,** A new method for preparation of phospholipid vesicles (liposomes) — French press, *FEBS Lett.,* 99, 210, 1979.

Chapter 11

MUSCARINIC RECEPTOR MECHANISMS IN ADRENAL CHROMAFFIN CELLS

Allan S. Schneider

TABLE OF CONTENTS

I. INTRODUCTION

Evidence for the existence of muscarinic acetylcholine receptors on adrenal chromaffin cells could be found as early as 1912 in a publication by Dale and Laidlaw,[1] in which pilocarpine, a muscarinic agonist, was shown to evoke adrenal catecholamine secretion. Although the actions of acetylcholine in the adrenal medulla were subsequently found to be predominantly nicotinic, Feldberg et al.[2] demonstrated in their classic study in 1934 the existence of "a subsidiary muscarinic action surviving nicotinic paralysis and requiring atropine for its suppression". Feldberg and co-workers[2] showed "that acetylcholine is the humoral transmitter of splanchnic impulses to the suprarenal medulla," and that "muscarine and pilocarpine cause an output of adrenaline from the suprarenals," the latter effect being completely abolished by atropine.

The ability of both nicotinic and muscarinic receptors to mediate adrenal catecholamine secretion in chromaffin cells of most animal species is now well documented (see Section VI). Activation of nicotinic receptors generally evokes a larger secretory response than that evoked by muscarinic receptors. The mechanism of nicotinic receptor initiation of secretion and the central role of calcium in "stimulus-secretion coupling" have been established in the pioneering studies of Douglas and co-workers.[3] They demonstrated receptor-mediated membrane depolarization leading to a key influx of calcium through voltage-gated calcium channels. In contrast, the mechanism by which muscarinic receptors evoke catecholamine release is presently unknown, although recent evidence suggests a different pathway for elevation of cytosolic calcium than that used by the nicotinic receptors.[4]

This chapter presents the current (as of 1985) state of knowledge of muscarinic receptor-mediated responses in adrenal chromaffin cells and their relation to possible mechanisms of catecholamine secretion. We begin with a discussion of the few binding studies demonstrating the presence and properties of muscarinic receptors on the chromaffin cell surface. This is followed by a review of several muscarinic receptor-mediated cellular responses, including elevation in cyclic GMP (cGMP), phosphoinositide metabolism, cytosolic calcium rise, and membrane calcium efflux. We next summarize what is known about muscarinic receptor-mediated catecholamine secretion, its calcium requirements, possible inhibitory effects, and preferential release of specific catecholamines. Finally, we attempt to provide a coherent view of the various muscarinic receptor-mediated responses, including their mutual effects upon one another and their possible relation to the secretory mechanism.

We shall not discuss the recent classification of muscarinic receptors into the putative M_1 and M_2 receptor subtypes,[5] since at this point in time there is no compelling evidence for such subtypes in the chromaffin cell and there is the possibility that the so-called M_1 and M_2 muscarinic receptor subtypes may be manifestations of different responses occurring at different levels of occupancy of a single receptor type.[6] We shall also ignore the large literature on muscarinic receptor responses in parasympathetic effector cells and nervous tissue, except where they are directly relevant to the discussion of chromaffin cell muscarinic receptor function.

II. PRESENCE OF MUSCARINIC RECEPTORS ON CHROMAFFIN CELLS — BINDING STUDIES

There have been only a few studies characterizing the radioligand binding properties of muscarinic receptors in the adrenal medulla. Following the technique of Yamamura and Snyder[7] for [^{3}H]-quinuclidinylbenzylate (^{3}H-QNB) binding to rat brain muscarinic receptors, Kayaalp and Neff[8,9] have demonstrated the presence of muscarinic receptors in both rat and bovine adrenal medulla. The rat medulla had about 20 times more binding sites per milligram protein than the bovine (66 fmol receptors per milligram protein — rat vs. 3 fmol receptors per milligram protein — bovine), which is interesting in light of the fact that the rat adrenal gland gives significant catecholamine release in response to muscarinic stimulation,[10] while the bovine gives little or none.[11] However, caution should be used in comparing receptor densities normalized to milligram protein for the two species, since the studies on the bovine were done with a crude membrane fraction, while those for the rat used homogenized adrenal medulla. The affinity of both rat and bovine chromaffin cell muscarinic receptors for (^{3}H-QNB) was about the same (K_D = 0.07 nM)[8,9] and close to that found in rat brain (K_D = 0.06 nM).[7] The density of muscarinic receptors in rat brain (65 fmol receptors per milligram tissue)[7] was similar to rat adrenal medulla.[8] Denervation of rat adrenal medulla did not alter its QNB binding properties, thus, confirming the location of the measured muscarinic receptors on chromaffin cells as opposed to the splanchnic nerve.[8]

A similar (^{3}H-QNB) binding study has been reported for transformed rat chromaffin cells in culture, i.e., for the PC12-pheochromocytoma cell line, including the effects of nerve growth factor on muscarinic receptor density.[12] High affinity QNB binding sites were found with a density of about 30 fmol/mg protein and a K_D of 0.15 nM — magnitudes similar to those for the normal chromaffin cell. Incubation of the PC12 cells for 12 to 15 days with nerve growth factor caused a marked increase in the density of muscarinic receptors from 30 to 157 fmol/mg protein.[12]

A comparison may be made between muscarinic and nicotinic receptor densities in the adrenal medulla. The latter have thus far only been measured by binding of [^{125}I]-α-bungarotoxin (BGTX) to plasma membrane fractions of bovine adrenal medulla.[13] Up to 190 fmol of BGTX binding sites per milligram membrane protein were found, which corresponded to 12 fmol/mg protein in a crude homogenate fraction. The relevance of the BGTX binding sites to functional nicotinic receptors on chromaffin cells is obscure, since BGTX does not inhibit nicotine-evoked catecholamine secretion in the adrenal medulla.[13,14]

III. cGMP RESPONSE

Cholinergic stimulation of adrenal chromaffin cells is known to induce an elevation of both cyclic AMP (cAMP) and cGMP.[15,16] Because the cAMP response is detectable only after about 5 to 10 min, it has generally been considered to be too slow to be involved in the secretory mechanism, which is activated on a time scale of seconds. Instead, the focus of interest in cAMP has been more in the area of catecholamine biosynthesis.[15] In contrast to the slow kinetics of cAMP elevation, we have found that bovine chromaffin cells exhibit a rapid rise (seconds) in cGMP upon cholinergic stimulation[16] — a rise whose kinetics were compatible with a possible involvement in the secretory mechanism (Figure 1). The chromaffin cell cGMP response was consistent with the finding of Aunis et al.[17] of guanylate cyclase activity in the plasma membrane fraction of bovine chromaffin cells. However, the dose-response curve for the ACh-induced rise in cGMP indicated a half-maximal response at an ACh concentration of 0.2 μM (Figure 1), which is almost two orders of magnitude lower than the ACh dose (10 μM) for half-maximal secretion.[16] The secretory response is known to be mediated predominantly by nicotinic ACh receptors[11] and the lower ACh concentrations required for elevation of cGMP indicated a muscarinic receptor-mediated response.[16] The muscarinic nature of the receptor mediating the rise in chromaffin cell cGMP was subsequently confirmed by the appearance of this cGMP rise in response to specific muscarinic agonists.[18]

Further characterization of the muscarinic-stimulated rise in bovine chromaffin cell cGMP included a report on its calcium requirements in comparison with those for ACh-evoked catecholamine secretion.[19] The stimulated rise in cGMP was shown to require a lower calcium concentration (0.2 mM) for half-maximal response than did catecholamine release (2.5 mM). Induced uptake of calcium by the calcium ionophore A23187 or by membrane depolarization

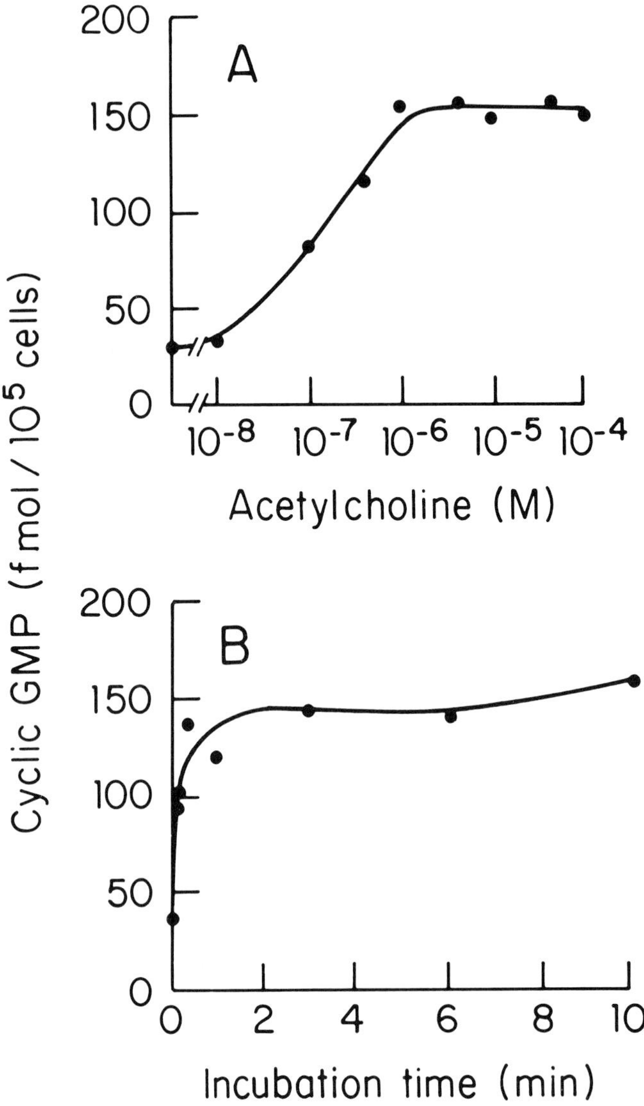

FIGURE 1.   Effect of acetylcholine on total cGMP in isolated bovine chromaffin cells. (A) Dose-response curve for ACh-stimulated cGMP accumulation. Incubation time 3 min; (B) kinetics of cGMP accumulation stimulated by 10 μM ACh.[16]

with excess potassium was reported by Lemaire et al.[19] to significantly increase chromaffin cell cGMP levels. However, the effect of excess K depolarization on chromaffin cell cGMP levels was not confirmed by Oka and co-workers,[18] and the concentration of A23187 (10 μM) used by Lemaire et al.[19] may have been toxic. Furthermore, neither membrane depolarization by veratridine nor the calcium ionophore X537A produced any effect on cGMP levels, although both evoked significant catecholamine secretion.[19] Thus, it appears that calcium uptake in itself may not be sufficient to evoke a rise in chromaffin cell cGMP. Removal of calcium from the extracellular medium was reported to block the ACh-evoked rise in cGMP, however, the calcium channel blockers, verapamil and D600, had only a marginal effect on the cGMP rise while blocking most of the secretory response.[19] Results obtained in calcium-free media are sometimes complicated by depletion of intracellular calcium pools, and caution should be used when interpreting such data. The effects of calcium on the cholinergic-stimulated cGMP rise and secretory response in bovine chromaffin cells are apparently manifested through different mechanisms — the former associated with the muscarinic receptor and the latter predominantly with the nicotinic receptor pathway. These two receptor pathways are now thought to mobilize calcium by different mechanisms,[4] and such differences may be what is being manifested in the different calcium dependence of the muscarinic-evoked rise in cGMP and the nicotinic-stimulated catecholamine secretion.

The muscarinic receptor-mediated rise in cGMP has been claimed to act as an inhibitory modulator of nicotine-evoked secretion in bovine chromaffin cells.[20] Addition of high concentrations of dibutyrl cGMP (up to 1 mM) or preincubation of the cells with muscarinic doses of ACh (0.2 to 0.5 μM) was reported to inhibit nicotine-evoked secretion. However, there have been conflicting reports on the effects of muscarinic agonists on nicotine-stimulated secretion from bovine chromaffin cells, and the inhibitory effect on secretion is now in question (see Section VI.C).

At present it is clear that chromaffin cell muscarinic receptors evoke a large (three- to fivefold) and rapid (seconds) increase in cGMP, however, the role of this cGMP increase in chromaffin cell function is still obscure.

IV. MUSCARINIC RECEPTOR-MEDIATED PHOSPHOINOSITIDE METABOLISM

A. Phosphoinositide Turnover

The first evidence of cholinergic receptor-stimulated phospholipid turnover in the adrenal medulla was provided by Hokin et al.[21] in one of a series of papers demonstrating a correlation between stimulated secretion and enhanced incorporation of ^{32}P into phosphatidic acid (PA) and phosphatidylinositol (PI) in several different secretory cell types.[21-23] They showed that acetylcholine stimulated the incorporation of ^{32}P into PI and PA with no significant effect on either phosphatidylcholine or phosphatidylethanolamine.[21] Hokin et al. provided dose-response curves for ACh-induced secretion and PI and PA labeling which showed a one to two order of magnitude lower ACh concentration for maximum phospholipid labeling than for maximum secretion. Although suggestive of a muscarinic receptor-mediated phospholipid turnover and nicotinic receptor-mediated secretion, the specific nature of the cholinergic receptors involved was not pursued in these early studies. An atropine block of both phospholipid labeling and secretion was also shown, however, the atropine concentration used (10 μM) was too high to clearly distinguish muscarinic and nicotinic receptor responses. Hokin et al.[21] concluded that the turnover of phospholipids plays a role in the secretion of small molecules, e.g., adrenaline from the adrenal medulla, in an analogous manner to its role in the secretion of proteins from exocrine glands and peptides from endocrine glands.

This work was extended by Trifaró, who first confirmed that ACh (10 μM) increased the incorporation of ^{32}P into PI and PA in bovine adrenal medullary slices and then showed that this was due to an increased turnover of both phosphatides rather than to increased uptake of label into the tissue or to increased specific activity of the nucleotide pool.[24] No difference was found in individual lipid contents upon ACh stimulation, however, the enhanced incorporation of ^{32}P into PA preceded that of PI, suggesting the former served as a precursor for synthesis of the latter. A 10-min lag time was observed before significant ACh-induced PA and PI labeling occurred, although stimulated secretion was evident much sooner.[24] It was also shown that omission of calcium abolished ACh-evoked secretion from slices of bovine adrenal medulla, but had no effect on ^{32}P incorporation into PI and PA.[25] Because of the differences in calcium dependence and kinetics, Trifaró concluded that there is no correlation between the ACh-induced PA and PI labeling and catecholamine secretion.[25]

There have been several recent reports demonstrating that the cholinergic-stimulated incorporation of ^{32}P into PI and PA in bovine chromaffin cells is mediated by muscarinic receptors.[26,27] Nicotinic ACh receptors were shown to evoke catecholamine release with little[26] or no[27] effect on PA and PI labeling. The lack of effect of calcium omission on PI and PA labeling, while blocking secretion, was confirmed.[26,28] Extreme calcium deprivation during a 15-min exposure to EGTA did inhibit phospholipid labeling,[26] however, this may be due to lowering of cytosolic calcium or to depletion of a relevant intracellular calcium pool. Stimulation of secretion (and presumably calcium uptake) by membrane depolarization with KCl was not accompanied by any change in PI or PA labeling.[28] Thus, it appears that omission of extracellular calcium has no effect on stimulated PI and PA labeling with ^{32}P and that calcium uptake does not enhance such incorporation. The main requirement for enhanced turnover of PI and PA in chromaffin cells appears to be muscarinic ACh receptor activation.

It is interesting that while all of the previous studies of ^{32}P labeling of PA and PI in adrenal medulla have reported an increased incorporation of radiolabel upon cholinergic stimulation,[21-27] Azila and Hawthorne[28] have seen a decrease. This may be due to the fact that they have washed out the excess ^{32}P following the initial incubation with label and have thus prevented continuing uptake of ^{32}P into the tissue during the course of their experiments. Thus, they are likely observing the initial breakdown of labeled PI, while other workers who maintain the label present throughout the stimulation experiment may be observing the resynthesis during the enhanced cycling of the phosphoinositides. Azila and Hawthorne[28] have also determined the subcellular localization of the phospholipid changes due to muscarinic receptor stimulation of perfused bovine adrenal medulla. Following stimulation of the prelabeled perfused gland with carbachol, subcellular fractions were prepared from the medulla. It was found that the decreased PI and PA labeling is not confined to the plasma membrane, but is distributed throughout the microsomal, granule, and mitochondrial membranes as well.[28]

The kinetics of the muscarinic-stimulated breakdown of PI and its subsequent resynthesis in freshly isolated bovine chromaffin cells have recently been measured in our laboratory. We have used a method of prelabeling isolated chromaffin cells with ^{32}P and removing the label prior to stimulation.[30] Figure 2 shows an initial drop in muscarine-stimulated PI labeling (2 min) followed by a gradual resynthesis of labeled PI over a period of 1 hr.[29] Lithium is known to inhibit the resynthesis of PI by preventing the formation of inositol from inositol phosphate. Addition of LiCl (10 mM) to the medium abolished the increased labeling of PI at 60 min following the initial decrease.[29]

Because some of the effects of receptor-induced turnover of phosphoinositides (e.g., intracellular calcium mobilization) are thought to be mediated by the breakdown of phosphatidylinositol-bisphosphate (PI-P$_2$) and the accumulation of inositoltriphosphate (IP$_3$), we have measured these polyphosphoinositides during muscarinic receptor stimulation of freshly isolated bovine chromaffin cells prelabeled with either ^{32}P or ^{3}H-inositol. Figure 3 shows that muscarine causes a rapid breakdown of PI-P$_2$ (measured as a loss of ^{32}P from the PI-P$_2$ band on TLC plates) within 2 min and that this effect is reversed by 0.5 μM atropine.[29] The corresponding accumulation of IP$_3$ during muscarinic stimulation of bovine chromaffin cells for 2 min and its reversal by atropine (0.5 μM) are shown in Figure 4. Thus, the muscarinic acetylcholine receptor on adrenal chromaffin cells regulates the turnover of mono- and polyphosphoinositides. The breakdown products of the phosphoinositide lipids, e.g., inositol sugars and diacylglycerol (DAG), are now thought to act as subcellular regulators of a variety of responses, including intracellular calcium mobilization by IP$_3$, activation of protein kinase C by DAG, and generation of arachidonic acid by hydrolysis of DAG or activation of phospholipase A. Some of these responses have recently begun to be seen in chromaffin cells.

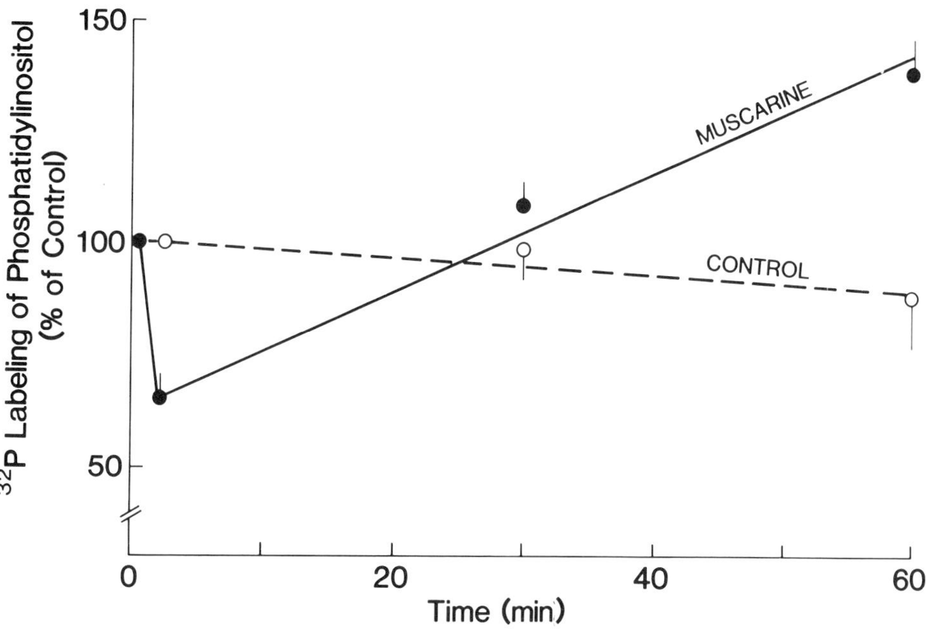

FIGURE 2.   Kinetics of muscarine-induced ^{32}P labeling of phosphatidylinositol in isolated bovine chromaffin cells. Cells were prelabeled with ^{32}P for 45 min, washed without ^{32}P, and incubated with or without muscarine (0.3 mM) for the indicated time period. Phospholipids were extracted and separated by thin layer chromatography.[29]

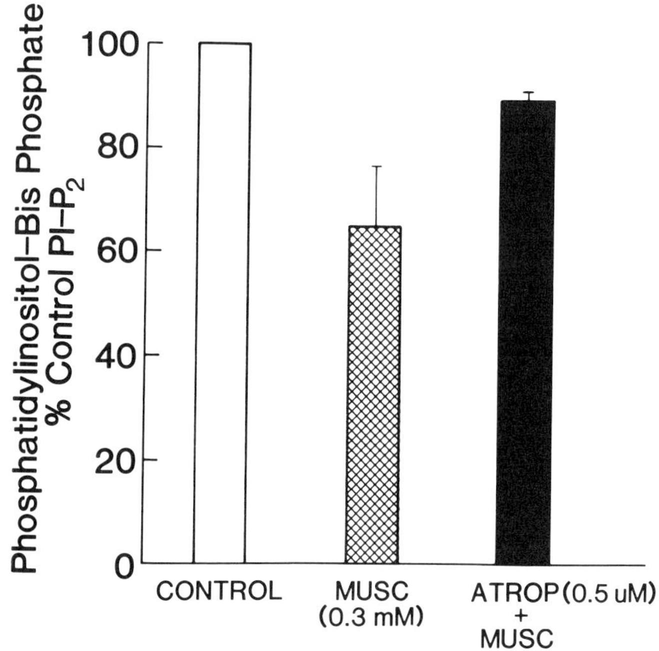

FIGURE 3.   Muscarine-stimulated phosphatidylinositol-bisphosphate (PI-P$_2$) breakdown. Cells were prelabeled for 45 min with ^{32}P, washed, and incubated with muscarine (0.3 mM) and LiCl (10 mM) in the presence or absence of atropine (0.5 μM). Phospholipids were extracted and separated by thin layer chromatography.[29]

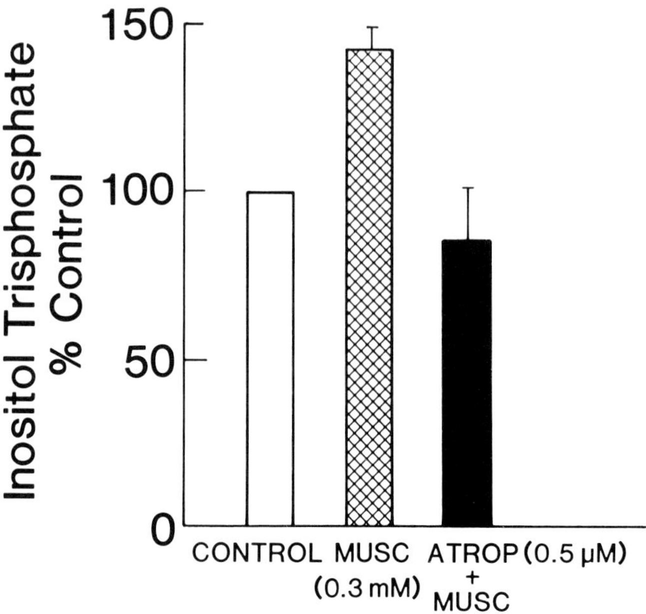

FIGURE 4.   Muscarine-stimulated accumulation of inositol trisphosphate in isolated bovine chromaffin cells. Cells were prelabeled with (^{3}H)-inositol for 3.5 hr, washed, and incubated for 2 min with muscarine (0.3 mM) and LiCl (10 mM) in the presence or absence of atropine (0.5 µM). The aqueous phase, obtained after extraction of phospholipids, was analyzed by anion exchange chromatography.[29]

B. Arachidonic Acid Generation

Arachidonic acid is one of a class of *cis*-unsaturated fatty acids that are potent membrane fusogens[31] and which can be generated by a calcium signal via phospholipase A$_2$ activation or by activation of DAG lipase. Addition of arachidonic acid to isolated bovine chromaffin cells has been reported to stimulate catecholamine secretion.[32] In the same study it was shown that an inhibitor of arachidonic acid metabolism, BW 755-c, a lipoxygenase inhibitor, blocked catecholamine release, while the cycloxygenase inhibitor, indomethasin, had no significant effect on release.[32] It has also been shown that arachidonic acid release accompanies catecholamine secretion from intact and permeabilized chromaffin cells.[33,34] The release of both arachidonic acid and catecholamines from intact bovine chromaffin cells was mediated by both nicotinic receptor activation and KCl-induced membrane depolarization, but not by muscarinic receptor agonists.[33] Considering that chromaffin cells from most species other than bovine give an easily measured secretory response to muscarinic receptor stimulation, it would be interesting to know whether they also give a muscarine-induced release of arachidonic acid.

In leaky chromaffin cells it was found that the cytosolic calcium dependence for arachidonic acid release was shifted to the left of that for catecholamine secretion, i.e., the former had a half-maximum response at about 0.4 µM and the latter at about 2 µM.[34] The kinetics of both responses were similar. The phospholipase A$_2$ inhibitor, *p*-bromophenacyl bromide, blocked both exocytosis and arachidonic acid release, which in the leaky cell could not be ascribed to its known effect as an inhibitor of calcium uptake in intact cells.[35,36] The phorbol ester, TPA, a known activator of protein kinase C, evoked both arachidonic acid release and exocytotic catecholamine release from the permeabilized cells.[34] Magnesium (10 to 20 mM) inhibited secretion without affecting arachidonic acid release from the leaky cells.

Because there is little increase in DAG in the leaky chromaffin cells, it was assumed that the release of arachidonic acid is stimulated via activation of phospholipsase A_2 rather than via DAG lipase.[34] However, since in the intact cell there is evidence for muscarinic-stimulated phosphoinositide lipid breakdown (see Figures 2 and 3 and Azila and Hawthorne[28]), there should also be accumulation of DAG. In addition, muscarinic receptors on bovine chromaffin cells have also been shown to mediate a rise in cytosolic calcium from about 100 to about 200 nM,[4] which, according to the calcium activation curve for arachidonic acid release in the leaky cell,[34] should be close to the levels required for evoking such release. However, the calcium levels reached may be marginal and no arachidonic acid release was detected in response to the muscarinic agonist, methacholine, in intact cultured bovine chromaffin cells.[33] A larger muscarinic response would be expected in chromaffin cells from other species and it may be useful to check for the DAG lipase pathway for arachidonic acid generation and release in such species.

C. Protein Kinase C Activation

The DAG generated during enhanced turnover of phosphoinositides is thought to be a potent activator of protein kinase C,[37] and a physiological pathway for such activation would be the muscarinic ACh receptors on adrenal chromaffin cells. The possible involvement of protein kinase C activation in the secretory mechanism in chromaffin cells was first suggested by the work of Knight and Baker with chromaffin cells permeabilized by exposure to high electric fields.[38] They showed that the calcium activation curve for exocytotic release of catecholamines is shifted to the left upon exposure of the leaky cells to the phorbol ester TPA. The implication of this work was that phorbol ester activation of protein kinase C sensitizes the exocytosis machinery to cytosolic calcium in the submicromolar range. Similar results were subsequently reported for digitonin permeabilized cells.[39] Consistent with these findings are recent results with intact chromaffin cells which exhibit a slow secretory response upon exposure to phorbol esters,[39,40] but which show a synergistic response when added in the presence of a calcium ionophore.[40] A similar synergism was observed in transformed PC12 pheochromocytoma cells: the calcium ionophore ionomycin gave a dose-response curve for release of catecholamine which was shifted to the left by addition of the phorbol ester TPA.[41] The observed potentiation of release occurred without any modification of the ionophore-induced rise in cytosolic calcium.[41]

The work cited above with phorbol ester activators of protein kinase C has generally given an enhancement of secretion. A recent report on the effects of the phorbol ester PMA on PC12 cells indicates an inhibitory effect on muscarinic receptor-mediated phosphoinositide turnover and cytosolic calcium rise.[42] Since the muscarinic receptor should mediate the activation of protein kinase C via phosphoinositide hydrolysis and DAG generation, the inhibitory effects of phorbol esters on muscarinic receptor responses suggest a possible negative feedback mechanism operating via activation of protein kinase C.[42]

There have been a number of other studies examining the phosphorylation of chromaffin cell proteins upon stimulation of the cell with secretagogues,[43-45] phorbol esters,[39,40] and cAMP and calcium.[46] These are beyond the scope of the present chapter and will not be considered further here. The reader is also referred to Chapter 6 for a discussion of calcium-calmodulin-dependent protein kinase activity in chromaffin cells.

V. MUSCARINIC RECEPTOR-MEDIATED CALCIUM MOBILIZATION

The central dogma of stimulus-secretion coupling is that there is a key requirement for a rise in cytosolic-free calcium for activation of exocytotic secretion.[47] In the adrenal chromaffin cell, the source of this rise has generally been considered to be a cholinergic receptor-mediated membrane depolarization accompanied by an uptake of extracellular calcium through voltage-gated calcium channels. This pathway for cholinergic receptor regulation of cytosolic calcium has been well documented for nicotinic ACh receptors on chromaffin cells.[3,47-51] The mechanism of muscarinic ACh receptor activation of catecholamine secretion from adrenal chromaffin cells, although somewhat more obscure, had also been thought to require an influx of extracellular calcium.[52] However, new evidence based on measurements of cytosolic-free calcium[4] and [45]Ca uptake and efflux during nicotinic and muscarinic stimulation of isolated chromaffin cells suggests that these receptors use different mechanisms for regulation of the rise in calcium during the secretory response.

A. Cytosolic-Free Calcium Response

Quin 2 is a fluorescent derivative of EGTA that has been developed as a probe of cytosolic-free calcium.[53] Its use in chromaffin cells has been demonstrated by several laboratories, where it was shown that ACh and other secretagogues evoke a rapid rise (seconds) in cytosolic calcium, from resting levels of about 100 nM to levels approaching the micromolar range.[4,54-58] It was discovered in our laboratory that a small component of the ACh-stimulated rise in cytosolic calcium is independent of extracellular calcium and is mediated by muscarinic receptors.[4] Figure 5 illustrates this component of the ACh-evoked rise in cytosolic calcium. It is seen that about a 50- to 100-nM cytosolic calcium rise is stimulated by ACh in calcium-free media (about 30 sec of 2 mM EGTA prior to addition of ACh) and that this rise is not affected by hexamethonium, but is blocked by 0.5 μM atropine. Further evidence of the muscarinic nature of the external calcium-independent rise in cytosolic calcium is shown in Figure 6. Muscarine (0.1 mM) is seen to evoke a small cytosolic calcium rise in calcium-free media and atropine totally blocks it. These findings have recently been confirmed in another laboratory using the muscarinic agonist methacholine.[55] We have further characterized the muscarinic receptor regulation of cytosolic calcium levels including appropriate dose-response curves for muscarine and ACh in calcium-containing and -free media, a comparison of the ability of various muscarinic agonists to evoke the response and the effect of muscarine on nicotinic receptor-mediated calcium transients.[58] The muscarinic agonoists methacholine and muscarine were found to be more effective than bethanechol, and pilocarpine was without effect on bovine chromaffin cells when all drugs were at 0.3 mM.

It may now be concluded that the nicotinic and muscarinic ACh receptors on adrenal chromaffin cells regulate cytosolic calcium by two independent pathways: the former via membrane depolarization and an influx of extracellular calcium leading to a cytosolic calcium rise into the micromolar range; the latter by mobilization of calcium from an intracellular source, via polyphosphoinositol turnover (PI-P$_2$ hydrolysis and IP$_3$ generation) leading to a small rise in cytosolic calcium to about 200 nM in bovine chromaffin cells. In chromaffin cells from other species with a greater muscarinic secretory response and a larger number of muscarinic receptors, the muscarine-evoked rise in cytosolic calcium might be expected to be somewhat larger.

B. [45]Ca Uptake and Efflux

Douglas and Poisner were the first to demonstrate an ACh-stimulated uptake of [45]Ca during activation of catecholamine secretion in the perfused adrenal medulla.[59] Subsequent studies with isolated and cultured bovine chromaffin cells have confirmed the fact that ACh stimulates a rapid uptake of [45]Ca concomitant with catecholamine release and that this uptake is mediated by nicotinic ACh receptors.[49,50,60] It was further shown that muscarinic ACh receptors did not induce [45]Ca uptake,[26,50,60] but instead evoked only an enhanced efflux.[60] The concentration of ACh evoking half-maximal uptake was 30 μM and that evoking half-maximal efflux was almost two orders of magnitude lower, consistent with the uptake being mediated by nicotinic receptors and the efflux by muscarinic receptors.[60] In addition, muscarine and ACh were shown to stimulate [45]Ca efflux[60,61] and nicotine and DMPP to stimulate

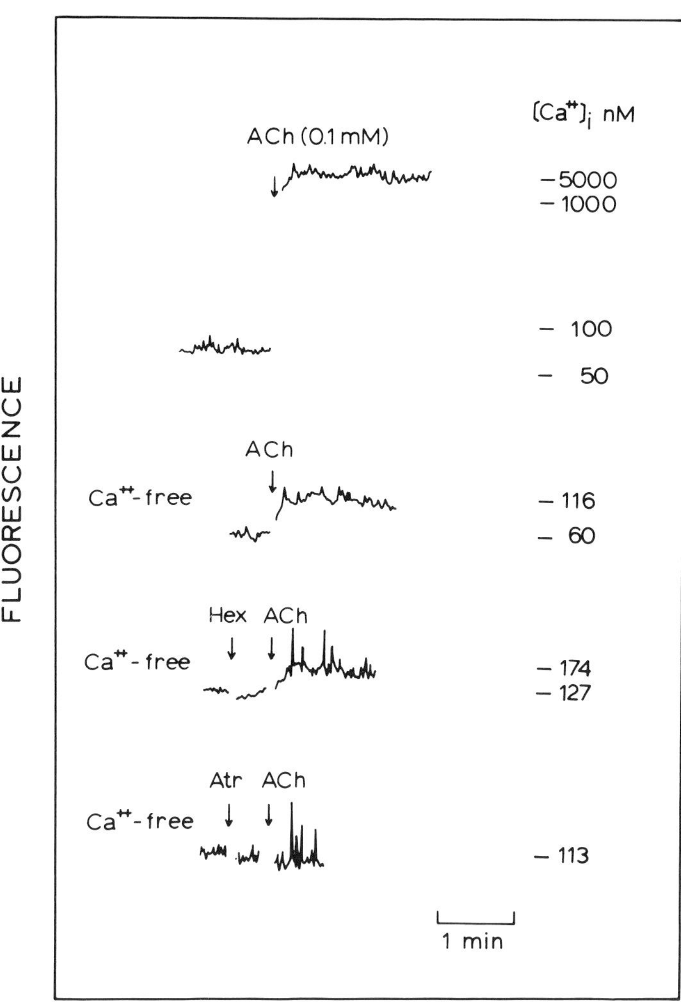

FIGURE 5.   Muscarinic nature and calcium dependence of the acetylcholine-stimulated rise in cytosolic calcium and quin 2 fluorescence in isolated bovine chromaffin cells. Cells were loaded with quin 2 and were resuspended in either calcium (2.2 mM)-containing buffer or calcium-free buffer containing 2 mM EGTA. Concentrations were ACh, 0.1 mM; hexamethonium, 0.5 mM; and atropine, 0.5 μM.[4]

^{45}Ca uptake.[49,50,60] The muscarinic agonists methacholine and muscarine had no significant effect on uptake, which when stimulated by ACh or carbachol was blocked by the nicotinic antagonists hexamethonium and mecamylamine.[50,60] Consistent with the lack of effect of muscarinic agonists on calcium uptake is the inability of 0.1 mM muscarine to cause any electrical depolarization of the plasma membranes of rat chromaffin cells.[62] Earlier electro-physiological measurements indicated a small depolarization of gerbil chromaffin cells induced by high doses of pilocarpine, however, these cells were apparently damaged since their resting potential was considerably below that shown in more recent studies and they did not exhibit action potentials.[63] There has been one report in which neither muscarine (0.1 mM) nor pilocarpine (0.1 mM) was able to elicit ^{45}Ca efflux from cultured bovine chromaffin cells.[64] Instead, it was reported that nicotine induced enhanced efflux following calcium uptake.

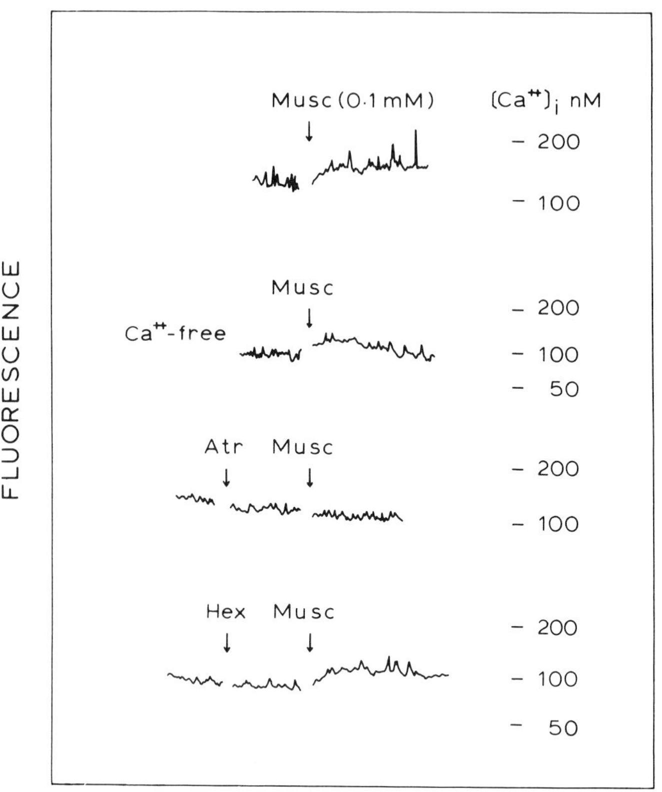

FIGURE 6.   Effects of muscarine of quin 2 fluorescence and cytosolic calcium in isolated bovine chromaffin cells in the presence and absence of extracellular calcium. Quin 2-loaded chromaffin cells were resuspended in either calcium-containing (2.2 mM) or calcium-free (2 mM EGTA) buffer and their fluorescence measured. Concentrations were muscarine (Musc), 0.1 mM; hexamethionium (Hex), 0.5 mM; atropine (Atr), 0.5 μM. Time scale same as in Figure 5.[4]

VI. CATECHOLAMINE SECRETION

There is substantial literature documenting a muscarinic component of cholinergic receptor-mediated catecholamine secretion from the adrenal medulla.[1-3,10,21,65-77] We have already mentioned in the Introduction the classic studies of Dale and Laidlaw[1] and Feldberg et al.[2] dating from the early part of the 20th century on the cat "suprarenal" gland. Subsequent studies of muscarinic ligand effects on adrenal catecholamine secretion were done on cat,[52,65-67,76,77] dog,[68-71,74] rat,[10,72] bovine,[11,13,16,18-20,26,60,78,79] guinea pig,[21,75] chicken,[73] and hamster[14] chromaffin cells. The early studies were done mainly on perfused cat and dog adrenal glands and the more recent work has emphasized isolated and cultured bovine, guinea pig, and hamster chromaffin cells. Four types of studies have been done: (1) those reporting effects of muscarinic receptor ligands on catecholamine release, (2) those examining the calcium requirements for muscarinic receptor-mediated secretion, (3) those examining the possible role of muscarinic receptors as inhibitory modulators of nicotine-evoked secretion, and (4) those examining the preferential release of epinephrine vs. norepinephrine by activation of muscarinic vs. nicotinic receptors. Each of these is reviewed briefly below.

A. Muscarinic Receptor-Stimulated Secretion

Muscarinic agonists have been found to evoke a significant release of catecholamines from the adrenal chromaffin cells of all animal species thus far examined, with the exceptions of bovine and hamster. The amount released is generally somewhat less than that evoked by activation of nicotinic receptors, perhaps about half as much.[10,75] In the one nonmammalian system thus far investigated, i.e., chick adrenals in organ culture, cholinergic-stimulated secretion has been reported to be purely muscarinic.[73]

The two species for which there have been reports of little or no muscarinic receptor-evoked secretion, i.e., bovine and hamster, have both been investigated mainly in the form of isolated or cultured cells, suggesting the possibility of muscarinic receptor damage during cell isolation. This seems unlikely since no significant methacholine- or pilocarpine-stimulated secretion has been found in perfused intact bovine adrenal medulla,[78] while a strong muscarinic receptor-mediated secretory response has been found with isolated guinea pig adrenal cells.[75] A plausible reason for the lack of significant secretion from bovine cells is the relatively low number of muscarinic receptors on this species[9] and the small rise in cytosolic calcium evoked by muscarinic agonists.[4,55] This rise in cytosolic calcium, from 100 to about 200 nM, is near the threshold of cytosolic calcium at which secretion begins to occur.[58,80] Careful examination in our laboratory of the specific effects of certain muscarinic agonists, i.e., muscarine and methacholine, on isolated and cultured bovine chromaffin cells indicates a small but reproducible secretory response amounting to about 3 or 4% of total cell catecholamines. In earlier studies from our laboratory we had shown a small muscarine-stimulated secretory response from isolated bovine chromaffin cells, but were unaware of its significance.[11] With the recent findings of a muscarinic pathway for raising cytosolic calcium in bovine cells that is distinct from the nicotinic pathway[4,55] and that utilizes the turnover of phosphoinositides,[23-29] the small bovine response has become more relevant. It is interesting that Role and Perlman[75] have also noted independent muscarinic and nicotinic secretory responses in isolated guinea pig adrenal cells based on additivity of maximal responses.

A word of caution is advised in interpreting some of the early atropine blocking data upon which claims of specific muscarinic effects were made. In many cases high, nonspecific, atropine concentrations were used. The concentration range in which atropine exerts a specific muscarinic block of chromaffin cell secretion has been shown to be below the micromolar range,[75] consistent with binding studies in rat brain.[7] Above micromolar concentrations, atropine begins to block chromaffin cell nicotinic receptors.[11,13,75,79]

B. Calcium Requirements of Muscarinic Receptor-Mediated Secretion

In spite of the large number of studies demonstrating muscarinic-induced adrenal catecholamine release and the central role of calcium in stimulus-secretion coupling, there have been only a few reports on the calcium requirements for such muscarine-induced secretion. Early studies on the perfused cat adrenal indicated that omission of calcium from the perfusion medium blocked catecholamine release evoked by muscarine, pilocarpine, and methacholine.[52] These studies utilized a 20-min perfusion with calcium-free media prior to adding muscarinic agonist. More recent studies have demonstrated a significant catecholamine secretory response evoked by pilocarpine and acetylcholine from cat adrenals perfused with calcium-free media containing 0.1 mM EGTA.[77] Secretagogues were added to the perfusion medium 25 min after exposure to calcium-free media and the amount released ranged up to a value of 25% of that in calcium-containing perfusion media. In contrast to the earlier study which suggested that muscarinic-evoked release utilized an influx of extracellular calcium, the more recent work demonstrated a significant muscarinic-evoked secretion that is independent of extracellular calcium and presumably utilizes an intracellular source.

Because of the possibility of muscarinic receptor mobilization of intracellular calcium as a mediator of secretion and the conflicting data on the perfused cat adrenal, we have recently reexamined the calcium requirements for muscarine- and nicotine-evoked catechomine release using perfused rat adrenal glands.[81] Some of our findings are shown in Figure 7. The

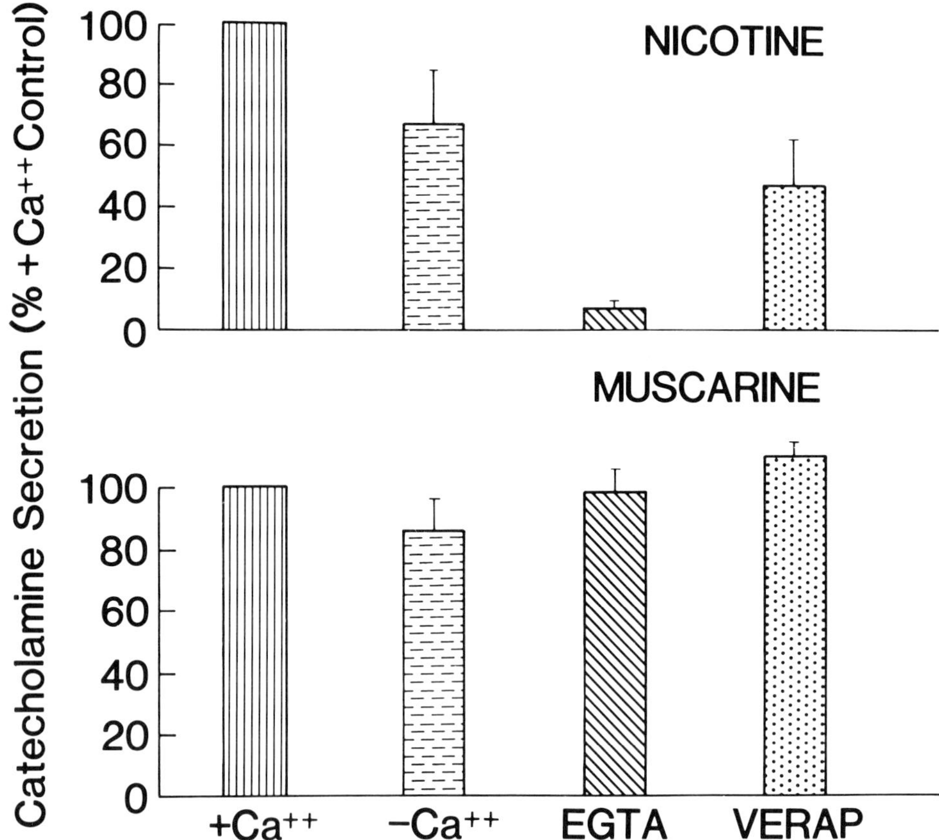

FIGURE 7.   Calcium requirements for nicotine-vs.-muscarine-evoked catecholamine secretion from perfused rat adrenal gland. The gland was perfused with either calcium-containing Krebs, calcium-free Krebs, calcium-free Krebs + 1 mM EGTA, or calcium-containing Krebs + verapamil (10 μM). For each perfusion condition catecholamine release was stimulated by nicotine injection (100 $\mu \ell$ of 7.5 μM) or muscarine injection (100 $\mu \ell$ of 2 to 5 mM).[81]

rat adrenal was perfused with either: (1) normal calcium (2 mM)-containing media, (2) normal perfusion media from which calcium was omitted, (3) calcium-free media containing 1 mM EGTA, or (4) normal perfusion media containing the calcium channel blocker, verapamil. The last three conditions had a clear inhibitory effect on nicotine-evoked secretion, while having no significant effect on muscarine-evoked secretion. The EGTA-containing perfusion medium totally blocked nicotine-evoked release within 1.0 min, while allowing muscarinic-evoked secretion to proceed unaltered. Longer periods of perfusion with EGTA-containing media (>5 min) caused a reduction in muscarine-evoked release, presumably due to depletion of an intracellular calcium pool used in the muscarine-stimulated response. These findings are consistent with previous studies of the calcium requirements of ACh-evoked secretion from perfused rat adrenals in which calcium-free media blocked transmural electrically stimulated release, but did not completely block secretion stimulated by acetylcholine.[82]

C. Muscarinic Inhibition of Nicotine-Evoked Secretion

The bovine adrenal gland provides an exception to the muscarinic receptor-mediated secretory response found in chromaffin cells from most other species. The nature of the bovine chromaffin cell response to muscarinic stimulation is at present controversial. There have been reports from two laboratories claiming a muscarinic inhibition of nicotine-evoked release in both perfused adrenal glands[78] and isolated chromaffin cells.[20] The degree of inhibition reported has not been large: 33% for low dose (0.5 μM) ACh inhibition of nicotine-evoked secretion from isolated chromaffin cells[20] and 43% for pilocarpine (0.1 mM) or methacholine (0.1 mM) inhibition of nicotine-evoked release from perfused glands.[78] At least two other laboratories have been unable to reproduce any specific muscarinic inhibition of nicotine-evoked release from isolated bovine chromaffin cells.[18,55] In one case 10 μM muscarine had no significant effect on the nicotine dose-response curve for catecholamine release. In a second report no methacholine inhibition of nicotine- or K-evoked secretion occurred at or below methacholine concentrations of 1 mM.[55] Higher metacholine concentrations (10 to 100 mM) did inhibit nicotine- and K-evoked release, but these effects were found to be nonspecific, since they were not reversed by atropine.[55] The reason for the above discrepancy between results from different laboratories is unclear. Our laboratory has searched for a muscarinic inhibition of nicotine-evoked secretion from isolated and cultured bovine chromaffin cells using muscarine, oxotremorine, and low dose ACh. Although it appeared from time to time that 1 μM muscarine, 0.5 μM oxotremorine, or 0.3 μM ACh was able to inhibit nicotine-evoked release from cultured chromaffin cells by about 30%, these results could not be consistently reproduced. At higher doses of muscarinic agonists the elusive inhibitory effect was lost and muscarine was found to enhance nicotine-evoked release. In freshly isolated bovine chromaffin cells we have been unable to detect any muscarinic inhibition of nicotinic responses.

There have been a few studies using chromaffin cells from nonbovine species in which muscarinic and nicotinic agonists were added together and their effects on secretion measured. For perfused cat adrenal glands, a range of muscarine doses (4.8 to 480 μM) failed to modify secretory responses to nicotine, high K, or veratridine.[76] For isolated guinea pig chromaffin cells, the effects of muscarine (0.2 mM) and nicotine (50 μM) were additive.[75]

D. Preferential Secretion of Epinephrine vs. Norepinephrine

Adrenal chromaffin cells are thought to be of two types: one containing mainly epinephrine and the other norepinephrine.[83,84] The preferential release of epinephrine by muscarinic agonists in perfused cat adrenals was first reported by Douglas and Poisner.[65] They found that muscarine and pilocarpine induced the release of mainly epinephrine (84 and 96% of total catecholamines, respectively), while addition of acetylcholine or nicotine to the perfusion medium caused large amounts of both epinephrine and norepinephrine to be released. The preferential release of epinephrine by muscarinic agonists was subsequently confirmed in the perfused cat[67] and rat[10] adrenal gland, although the preferential effect in the rat gland was relatively slight. The preferential release of epinephrine by muscarinic agonists was not found in the dog adrenal gland. Since adrenocorticosteroids are known to stimulate the conversion of norepinephrine to epinephrine, it would be interesting to determine in the cat adrenal medulla whether they also increase the proportion of muscarinic receptors present.

The fact that α- and β-adrenergic receptors are present in different proportions in different target tissues and can elicit different responses suggests a possible physiological significance for the preferential release of epinephrine vs. norepinephrine from the adrenal medulla. This could occur by selective innervation and control of epinephrine vs. norepinephrine containing chromaffin cells and/or by variation in the frequency of impulses traveling down the splanchnic nerve with consequent variation in ACh output. Lower levels of ACh release would selectively activate muscarinic receptors, since these have a higher affinity for ACh. The existence of a muscarinic receptor-regulated preferential release of catecholamines in chromaffin cells from species other than the cat and the physiological relevance of such a preferential release remain to be demonstrated.

VII. SUMMARY AND FUTURE DIRECTIONS

During recent years data have begun to accumulate on a new muscarinic receptor mechanism in chromaffin cells, a mechanism that is distinct from the well-established nicotinic receptor pathway and which may have implications for the mechanism of exocytotic secretion of catecholamines. Although the data are still quite scanty, a consistent picture has begun to emerge based on measurements of muscarinic receptor-mediated phosphoinositide turnover, cytosolic calcium responses, ^{45}Ca uptake and efflux, and catecholamine secretion. It appears that activation of muscarinic receptors on adrenal chromaffin cells causes the hydrolysis of phosphoinositide lipids and the generation of inositol trisphosphate (IP$_3$) and diacylglycerol.[21,24-29] The IP$_3$, in turn, can mobilize calcium from intracellular stores to raise cytosolic calcium in a manner initially independent of extracellular calcium.[4,55] As a result of the rise in intracellular calcium there is an enhanced efflux of calcium from the cytosol to the extracellular medium,[60] without an accompanying enhanced uptake of extracellular calcium.[26,50,60] In the bovine chromaffin cell this mobilization of cytosolic calcium is not sufficient to evoke a pronounced secretory response.[4,11,13,55] In chromaffin cells of other species, where larger muscarinic-evoked secretory responses are known to occur (see Section VI.A), one might expect intracellular calcium mobilization to result in higher levels of cytosolic calcium. Future experiments should clarify this question.

The generation of diacylglycerol is potentially interesting as both an activator of protein kinase C and a source of membrane fusogens. Activation of protein kinase C is believed to be the mechanism by which phorbol esters increase the sensitivity of the secretory response in leaky chromaffin cells to cytosolic calcium.[38,39] It would be interesting to know more about the identity, subcellular localization, and functions of the substrates of protein kinase C in chromaffin cells and their possible relevance to the secretory mechanism. Of the many chromaffin cell substrates that appear as phosphorylated proteins on gels, tyrosine hydroxylase is the only one that can be identified with a known function.[39,44,46] Despite the known actions of phorbol esters on chromaffin cells, evidence of a physiological activation of protein kinase C by muscarinic receptors in this system remains to be demonstrated.

An increase in the level of diacylglycerol may facilitate exocytotic fusion of chromaffin granules and plasma membranes. Diacylglycerol may itself be a fusogen. It is also a potential source of arachidonic acid via its breakdown by diacylglycerol lipase. Arachidonic acid may also derive from activation of phospholipase A, a calcium-requiring enzyme. Arachidonic acid is known to enhance fusion of chromaffin granule membranes.[31]

The muscarinic receptors on chromaffin cells have begun to be recognized as a transducer of a number of interesting cellular responses involving phosphoinositide turnover and calcium mobilization, with possible relevance to the mechanism of exocytotic secretion. The fact that chromaffin cell muscarinic receptors have a higher binding affinity for acetylcholine than do nicotinic receptors indicates that these muscarinic receptors will be activated under most physiological conditions of adrenal medullary secretion. Also relevant is the fact that the kinetics of the muscarine-induced rise in cytosolic calcium and phosphoinositide turnover is compatible with the kinetics of catecholamine secretion. Although some of the mechanisms suggested above are highly speculative as of this writing (fall 1985), they are amenable to experimental test, and new insights should be forthcoming with the rapid movement of this field.

REFERENCES

1. **Dale, H. H. and Laidlaw, P. O.,** The significance of the suprarenal capsules in the actions of certain alkaloids, *J. Physiol. (London),* 45, 1, 1912.
2. **Feldberg, W., Mintz, B., and Tsudzimura, H.,** The mechanism of nervous discharge of adrenaline, *J. Physiol. (London),* 81, 286, 1934.
3. **Douglas, W. W.,** Secretomotor control of adrenal medullary secretion: synaptic, membrane and ionic events in stimulus-secretion coupling, in *Handbook of Physiology,* Sect. 7, Vol. 6, Blaschko, H., Sayers, G., and Smith, A. D., Eds., American Physiological Society, Washington, D.C., 1975, chap. 26.
4. **Kao, L.-S. and Schneider, A. S.,** Muscarinic receptors on bovine chromaffin cells mediate a rise in cytosolic calcium that is independent of extracellular calcium, *J. Biol. Chem.,* 260, 2019, 1985.
5. **Vickroy, T. W., Watson, M., Yamamura, H. I., and Roeske, W. R.,** Agonist binding to multiple muscarinic receptors, *Fed. Proc.,* 43, 2785, 1984.
6. **Brown, J. H. and Masters, S. B.,** Muscarinic regulation of phosphatidylinositol turnover and cyclic nucleotide metabolism in the heart, *Fed. Proc.,* 43, 2613, 1984.
7. **Yamamura, H. I. and Snyder, S. H.,** Muscarinic cholinergic binding in rat brain, *Proc. Natl. Acad. Sci. U.S.A.,* 71, 1725, 1974.
8. **Kayaalp, S. O. and Neff, N. H.,** Muscarinic receptor binding in rat adrenal medulla, *Eur. J. Pharmacol.,* 57, 255, 1979.
9. **Kayaalp, S. O. and Neff, N. F.,** Cholinergic muscarinic receptors of bovine adrenal medulla, *Neuropharmacology,* 18, 909, 1979.
10. **Wakade, A. R. and Wakade, T. D.,** Contribution of nicotinic and muscarinic receptors in the secretion of catecholamines evoked by endogenous and exogenous acetylcholine, *Neuroscience,* 10, 973, 1979.
11. **Schneider, A. S., Herz, R., and Rosenheck, K.,** Stimulus-secretion coupling in chromaffin cells isolated from bovine adrenal medulla, *Proc. Natl. Acad. Sci. U.S.A.,* 74, 5036, 1977.
12. **Jumblatt, J. E. and Tishler, A. S.,** Regulation of muscarinic ligand binding sites by nerve growth factor in PC12 phaeochromocytoma cells, *Nature (London),* 297, 152, 1982.
13. **Wilson, S. P. and Kirshner, N.,** The acetylcholine receptor of the adrenal medulla, *J. Neurochem.,* 28, 687, 1977.
14. **Liang, B. T. and Perlman, R. L.,** Catecholamine secretion by hamster adrenal cells, *J. Neurochem.,* 32, 927, 1979.
15. **Guidotti, A., Hanbauer, I., and Costa, E.,** Role of cyclic nucleotides in the induction of tyrosine hydroxylase, *Adv. Cyclic Nucleotide Res.,* 5, 619, 1975.
16. **Schneider, A. S., Cline, H. T., and Lemaire, S.,** Rapid rise in cyclic GMP accompanies catecholamine secretion in suspensions of adrenal chromaffin cells, *Life Sci.,* 24, 1389, 1979.
17. **Aunis, D., Pescheloche, M., and Zwiller, J.,** Guanylate cyclase from bovine adrenal medulla: subcellular localization and studies of the effect of lysolecithin on enzyme activity, *Neuroscience,* 3, 83, 1978.
18. **Yanagihara, N., Isosaki, M., Ohuchi, T., and Oka, M.,** Muscarinic receptor-mediated increase in cyclic GMP levels in isolated bovine adrenal medullary cells, *FEBS Lett.,* 105, 296, 1979.
19. **Lemaire, S., Derome, G., Tseng, R., Mercier, P., and Lemaire, I.,** Distinct regulations by calcium of cyclic GMP levels and catecholamine secretion in isolated bovine adrenal chromaffin cells, *Metabolism,* 30, 462, 1981.
20. **Derome, G., Tseng, R., Mercier, P., Lemaire, I., and Lemaire, S.,** Possible muscarinic regulation of catecholamine secretion mediated by cyclic GMP in isolated bovine adrenal chromaffin cells, *Biochem. Pharmacol.,* 30, 855, 1981.
21. **Hokin, M. R., Benfy, B. G., and Hokin, L. E.,** Phospholipides and adrenal secretion in guinea pig adrenal medulla, *J. Biol. Chem.,* 233, 814, 1958.
22. **Hokin, L. E. and Hokin, M. R.,** Phosphoinositides and protein secretion in pancreas slices, *J. Biol. Chem.,* 233, 805, 1958.
23. **Hokin, M. R., Hokin, L. E., Saffran, M., Schally, A. V., and Zimmerman, B. U.,** Phospholipides and the secretion of adrenocorticotropin and of corticosteroids, *J. Biol. Chem.,* 233, 811, 1958.
24. **Trifaró, J. M.,** Phospholipid metabolism and adrenal medullary activity, *Mol. Pharmacol.,* 5, 382, 1969.
25. **Trifaró, J. M.,** The effect of calcium omission on the secretion of catecholamines and the incorporation of ^{32}P into nucleotides and phospholipides of bovine adrenal medulla during acetylcholine stimulation, *Mol. Pharmacol.,* 5, 420, 1969.
26. **Fisher, S. K., Holz, R. W., and Agranoff, B. W.,** Muscarinic receptors on chromaffin cell cultures mediate enhanced phospholipid labeling but not catecholamine secretion, *J. Neurochem.,* 37, 491, 1981.
27. **Mohd Adnan, N. A. and Hawthorne, J. N.,** Phosphatidylinositol labeling in response to activation of muscarinic receptors in bovine adrenal medulla, *J. Neurochem.,* 36, 1858, 1981.
28. **Azila, N. and Hawthorne, J. N.,** Subcellular localization of phospholipid changes in response to muscarinic stimulation of perfused bovine adrenal medulla, *Biochem. J.,* 204, 291, 1982.

29. **Harish, O. and Schneider, A. S.,** unpublished data.
30. **Rebecchi, M. J., Kolesnick, R. N., and Gershengorn, M. C.,** Thyrotropin releasing hormone stimulates rapid loss of phosphatidylinositol and its conversion to 1,2-diacylglycerol and phosphatidic acid in rat mammotropic pituitary cells. Association with calcium mobilization and prolactin secretion, *J. Biol. Chem.,* 258, 227, 1983.
31. **Creutz, C. E.,** *cis*-Unsaturated fatty acids induce the fusion of chromaffin granules aggregated by synexin, *J. Cell Biol.,* 91, 247, 1981.
32. **Nishibe, S., Ogawa, M., Murata, A., Nakamura, K., Hatanaka, T., Kambayashi, J., and Kosaki, G.,** Inhibition of catecholamine release from isolated bovine adrenal medullary cells by various inhibitors: possible involvement of protease, calmodulin and arachidonic acid, *Life Sci.,* 32, 1613, 1983.
33. **Frye, R. A. and Holz, R. W.,** The relation between arachidonic acid release and catecholamine secretion from cultured bovine adrenal chromaffin cells, *J. Neurochem.,* 43, 146, 1984.
34. **Frye, R. A. and Holz, R. W.,** Arachidonic acid release and catecholamine secretion from digitonin treated chromaffin cells: effects of micromolar calcium, phorbol esters and protein alkylating agents, *J. Neurochem.,* 44, 265, 1985.
35. **Wada, A., Sakurai, S., Kobayashi, H., Yanagihara, N., and Izumi, F.,** Suppression by phospholipase A_2 inhibitors of secretion of catecholamines from isolated adrenal medullary cells by suppression of cellular calcium uptake, *Biochem. Pharmacol.,* 32, 1175, 1983.
36. **Frye, R. A. and Holz, R. W.,** Phospholipase A_2 inhibitors block catecholamine secretion and calcium uptake in cultured bovine adrenal medullary cells, *Mol. Pharmacol.,* 23, 547, 1983.
37. **Nishizuka, Y.,** Turnover of inositol phospholipids in signal transduction, *Science,* 225, 1365, 1984.
38. **Knight, D. E. and Baker, P. F.,** The phorbol ester TPA increases the affinity of exocytosis for calcium in leaky adrenal medullary cells, *FEBS Lett.,* 160, 98, 1983.
39. **Pocotte, S. L., Frye, R. A., Senter, R. A., Ter Bush, D. R., Lee, S. A., and Holz, R. W.,** Effects of phorbol ester on catecholamine secretion and protein phosphorylation, *Proc. Natl. Acad. Sci. U.S.A.,* 82, 930, 1985.
40. **Brocklehurst, K. W., Morita, K., and Pollard, H. B.,** Characterization of protein kinase C and its role in catecholamine secretion from bovine adrenal medullary cells, *Biochem. J.,* 228, 35, 1985.
41. **Pozzan, T., Gatti, G., Dozio, N., Vicentini, L. M., and Meldolisi, J.,** Calcium dependent and independent release of neurotransmitter from PC12 cells: a role for activation of protein kinase C, *J. Cell Biol.,* 99, 628, 1984.
42. **Vicentini, L. M., DiVirgilio, F., Ambrosini, A., Pozzan, T., and Meldolisi, J.,** Tumor promoter, phorbol 12-myristate-13-acetate inhibits phosphoinositide hydrolysis and cytosolic calcium rise induced by activation of muscarinic receptors in PC12 cells, *Biochem. Biophys. Res. Commun.,* 127, 310, 1985.
43. **Amy, C. M. and Kirshner, N.,** Phosphorylation of adrenal medulla cell proteins in conjunction with stimulation of catecholamine secretion, *J. Neurochem.,* 36, 847, 1981.
44. **Haycock, J. W., Meligeni, J. A., Bennett, W. F., and Waymire, J. C.,** Phosphorylation and activation of tyrosine hydroxylase mediate acetylcholine-induced increase in catecholamine biosynthesis in adrenal chromaffin cells, *J. Biol. Chem.,* 257, 12641, 1982.
45. **Holz, R. W., Rothwell, G. E., and Ueda, T.,** Cholinergic agonist stimulated phosphorylation of two specific proteins in bovine chromaffin cells: correlation with catecholamine secretion, *Soc. Neurosci.,* 6, 177, 1980.
46. **Niggli, V., Knight, D. E., Baker, P. F., Vigny, A., and Henry, J. P.,** Tyrosine hydroxylase in leaky adrenal medullary cells: evidence for in-situ phosphorylation by separate Ca^{2+} and cyclic AMP-dependent systems, *J. Neurochem.,* 43, 646, 1984.
47. **Douglas, W. W.,** Stimulus secretion coupling: variations on the theme of calcium activated exocytosis involving cellular and extracellular sources of calcium, in *CIBA Foundation Symp. 54 (New Series): Respiratory Tract Mucus,* Elsevier, Amsterdam, 1978, 61.
48. **Douglas, W. W. and Rubin, R. P.,** The role of calcium in the secretory response of the adrenal medulla to acetylcholine, *J. Physiol. (London),* 159, 40, 1961.
49. **Kilpatrick, D. L., Slepetis, R. J., Corcoran, J. J., and Kirshner, N.,** Calcium uptake and catecholamine secretion by cultured bovine adrenal medulla cells, *J. Neurochem.,* 38, 427, 1982.
50. **Holz, R. W., Senter, R. A., and Frye, R. A.,** Relationship between Ca^{2+} uptake and catecholamine secretion in primary dissociated cultures of adrenal medulla, *J. Neurochem.,* 39, 635, 1982.
51. **Schneider, A. S., Cline, H. T., Rosenheck, K., and Sonenberg, M.,** Stimulus-secretion coupling in isolated adrenal chromaffin cells: calcium channel activation and possible role of cytoskeletal elements, *J. Neurochem.,* 37, 567, 1981.
52. **Poisner, A. M. and Douglas, W. W.,** The need of calcium in adrenomedullary secretion evoked by biogenic amines, polypeptides, and muscarinic agents, *Proc. Soc. Exp. Biol. Med.,* 123, 62, 1966.
53. **Tsien, R. Y., Pozzan, T., and Rink, T. J.,** Calcium homeostasis in intact lymphocytes: cytoplasmic free calcium monitored with a new intracellulary trapped fluorescent indicator, *J. Cell Biol.,* 94, 325, 1982.

54. **Knight, D. E. and Kesteven, N. T.,** Evoked transient intracellular free Ca^{2+} changes and secretion in isolated bovine adrenal medullary cells, *Proc. R. Soc. London Ser. B,* 218, 177, 1983.
55. **Cheek, T. R. and Burgoyne, R. D.,** Effect of activation of muscarinic receptors on intracellular free calcium and secretion in bovine adrenal chromaffin cells, *Biochim. Biophys. Acta,* 846, 167, 1985.
56. **Kao, L.-S. and Schneider, A. S.,** Quin 2 fluorescence measurements of cytosolic calcium during stimulus-secretion coupling in adrenal chromaffin cells, *Fed. Proc.,* 43, 770, 1984.
57. **Burgoyne, R. D.,** The relationship between secretion and intracellular free calcium in bovine adrenal chromaffin cells, *Biosci. Rep.,* 4, 605, 1984.
58. **Kao, L.-S. and Schneider, A. S.,** Calcium mobilization and catecholamine secretion in adrenal chromaffin cells: a quin 2 fluorescence study, *J. Biol. Chem.,* 261, 4881, 1986.
59. **Douglas, W. W. and Poisner, A. M.,** On the mode of action of acetylcholine in evoking adrenomedullary secretion: increased uptake of calcium during the secretory response, *J. Physiol. (London),* 162, 385, 1962.
60. **Oka, M., Isosaki, M., and Watanabe, J.,** Calcium flux and catecholamine release in isolated bovine adrenal medullary cells: effects of nicotinic and muscarinic stimulation, *Adv. Biosci.,* 36, 29, 1982.
61. **Ohsako, S. and Deguchi, T.,** Phosphatidic acid mimics the muscarinic action of acetylcholine in cultured bovine chromaffin cells, *FEBS Lett.,* 152, 62, 1983.
62. **Kidokoro, Y., Miyazaki, S., and Ozawa, S.,** Acetylcholine-induced membrane depolarization and potential fluctuations in the rat adrenal chromaffin cell, *J. Physiol. (London),* 324, 203, 1982.
63. **Douglas, W. W., Kanno, T., and Samson, S. R.,** Effects of acetylcholine and other medullary secretagogues and antagonists on the membrane potential of adrenal chromaffin cells: an analysis employing techniques of tissue culture, *J. Physiol. (London),* 188, 107, 1967.
64. **Wada, A., Izumi, F., Yashima, N., Kobayashi, H., Toyohira, Y., and Yanagihara, N.,** Efflux of ^{45}Ca from bovine adrenal medulla cells in culture caused by nicotinic receptor stimulation, *Neurosci. Lett.,* 47, 69, 1984.
65. **Douglas, W. W. and Poisner, A. M.,** Preferential release of adrenaline from the adrenal medulla by muscarine and pilocarpine, *Nature (London),* 208, 1102, 1965.
66. **Lee, F.-L. and Trendelenburg, U.,** Muscarinic transmission of preganglionic impulses to the adrenal medulla of the cat, *J. Pharmacol. Exp. Ther.,* 158, 73, 1967.
67. **Rubin, R. P. and Miele, E.,** A study of the differential secretion of epinephrine and norepinephrine from the perfused cat adrenal gland, *J. Pharmacol. Exp. Ther.,* 164, 115, 1968.
68. **Kayaalp, S. O. and McIsaac, R. J.,** In vivo release of catecholamines from the adrenal medulla by selective activation of cholinergic receptors, *Arch. Int. Pharmacodyn. Ther.,* 176, 168, 1968.
69. **Kayaalp, S. O. and McIsaac, R. J.,** Muscarinic component of splanchnic-adrenal transmission in the dog, *Br. J. Pharmacol.,* 36, 286, 1969.
70. **Kayaalp, S. O. and Turker, R. K.,** Evidence for muscarinic receptors in the adrenal medulla of the dog, *Br. J. Pharmacol.,* 35, 265, 1969.
71. **Kovacic, B. and Robinson, R. L.,** Drug-induced secretion of catecholamines by the perfused adrenal gland of the dog during nicotinic blockade, *J. Pharmacol. Exp. Ther.,* 175, 178, 1970.
72. **Yoshizaki, T.,** Participation of muscarinic receptors in splanchnic-adrenal transmission in the rat, *Jpn. J. Pharmacol.,* 23, 813, 1973.
73. **Ledbetter, F. H. and Kirshner, N.,** Studies of chick adrenal medulla in organ culture, *Biochem. Pharmacol.,* 24, 967, 1975.
74. **Tsujimoto, A. and Nishikawa, T.,** Further evidence for nicotinic and muscarinic receptors and their interaction in dog adrenal medulla, *Eur. J. Pharmacol.,* 34, 337, 1975.
75. **Role, L. W. and Perlman, R. L.,** Both nicotinic and muscarinic receptors mediate catecholamine secretion by isolated guinea pig chromaffin cells, *Neuroscience,* 10, 979, 1983.
76. **Kirpekar, S. M., Prat, J. C., and Schiavone, M. T.,** Effect of muscarine on release of catecholamines from the perfused adrenal glands of the cat, *Br. J. Pharmacol.,* 77, 455, 1982.
77. **Nakazato, Y., Yamada, Y., Tomita, U., and Ogha, A.,** Muscarinic agonists release adrenal catecholamines by mobilizing intracellular Ca^{2+}, *Proc. Jpn. Acad. B,* 60, 314, 1984.
78. **Swilem, A.-M. F., Hawthorne, J. N., and Azila, N.,** Catecholamine secretion by perfused bovine adrenal medulla in response to nicotinic activation is inhibited by muscarinic receptors, *Biochem. Pharmacol.,* 32, 3873, 1983.
79. **Trifaró, J. M. and Lee, R. W. H.,** Morphological characteristics and stimulus secretion coupling in bovine adrenal cell cultures, *Neuroscience,* 5, 1533, 1980.
80. **Knight, D. E. and Baker, P. F.,** Calcium-dependence of catecholamine release from bovine adrenal medullary cells after exposure to intense electric fields, *J. Membr. Biol.,* 68, 107, 1982.
81. **Harish, O. E., Kao, L.-S., Raffaniello, R., Wakade, A. R., and Schneider, A. S.,** Calcium-dependence of muscarinic receptor-mediated catecholamine secretion from the perfused rat adrenal medulla, *J. Neurochem.,* in press.
82. **Wakade, A. R.,** Studies on secretion of catecholamines evoked by acetylcholine or transmural stimulation of the rat adrenal gland, *J. Physiol. (London),* 313, 463, 1981.

83. **Coupland, R. E.,** *The Natural History of the Chromaffin Cell,* Longman Green, London, 1965.
84. **Hillarp, N. A. and Hokfelt, B.,** Evidence of adrenaline and noradrenaline in separate adrenal medullary cells, *Acta Physiol. Scand.,* 30, 55, 1953.

Chapter 12

SODIUM AND CALCIUM CHANNELS IN CULTURED BOVINE ADRENAL MEDULLA CELLS

Norman Kirshner

TABLE OF CONTENTS

I. INTRODUCTION

The development of methods for isolating and maintaining adrenal medullary chromaffin cells in culture[1-8] has led to a rapid expansion of our knowledge of this tissue. This chapter describes studies of the roles of sodium and calcium channels in stimulus-secretion coupling in cultured adrenal medullary cells.

II. SODIUM CHANNELS

A. Biochemical and Pharmacological Analysis

Neurotoxins have been valuable tools in examining the roles of ion channels in adrenal medullary cell cultures. The sites at which several neurotoxins act on voltage-sensitive sodium channels have been identified.[9] Tetrodotoxin (TTX) and saxitoxin bind at site I and inhibit ion flux through the sodium channel. Veratridine, batrachotoxin (BTX), aconitine, and grayanotoxin (GTX) bind at site II and cause persistent activation of the sodium channel, resulting in increased influx of NA^+. Scorpion toxin (*Leirus quinquestriatus*) and sea anemone toxin bind at site III and inhibit inactivation of the sodium channel and markedly enhance the activation caused by veratridine, BTX, and aconitine. The enhanced Na^+ flux induced by veratridine, BTX, aconitine, GTX, and scorpion venom is blocked by TTX. Additionally, histrionicotoxin (HTX), a neurotoxin derived from the skin of the Columbian frog *Dendrobates histrionicus*, binds to a site in the ion channel of the nicotinic receptor and blocks ionic currents in a reversible, noncompetitive manner.[10,11]

Batrachotoxin, veratridine, aconitine, and scorpion venom stimulate catecholamine secretion and promote influx of $^{22}Na^+$ in cultured chromaffin cells in a dose-dependent manner (Figure 1) which is inhibited by TTX.[12] The dose dependency for catecholamine secretion and $^{22}Na^+$ influx, as well as the dose dependency for inhibition of these processes by TTX, is superimposable (Figure 2). In neuroblastoma cells[13] scorpion toxin interacts cooperatively with BTX and veratridine to stimulate uptake of $^{22}Na^+$, and it has been shown that chromaffin cells react in a similar manner (Figure 3).[12] In adrenal medulla cells a concentration of scorpion venom (0.5 $\mu g/m\ell$) which has only a modest effect on catecholamine secretion and $^{22}Na^+$ uptake enhanced catecholamine secretion and $^{22}Na^+$ uptake induced by low concentrations of BTX or veratridine threefold. Scorpion venom reduced the concentrations needed to stimulate half-maximal $^{22}Na^+$ uptake from 20 μM to less than 5 μM for veratridine and from 0.3 μM to approximately 0.1 μM for BTX, but did not increase $^{22}Na^+$ influx above the levels obtained with saturating concentrations of either of the two alkaloids. Scorpion venom had the same effect on catecholamine secretion.

Nicotine also stimulates catecholamine secretion and $^{22}Na^+$ uptake in adrenal medullary cells. At concentrations of nicotine or K^+ which elicit less than 50% of the maximal secretory response, TTX partially inhibits catecholamine secretion.[14] However, at higher concentrations of nicotine catecholamine secretion and $^{22}Na^+$ uptake are not inhibited by TTX, but both are inhibited by HTX.[12] HTX has no effect on $^{22}Na^+$ influx induced by veratridine or batrachotoxin. Thus, while voltage-sensitive sodium channels are utilized in veratridine- and BTX-induced catecholamine secretion and $^{22}Na^+$ uptake, they are not obligatory for nicotinic receptor-mediated catecholamine secretion or $^{22}Na^+$ uptake (Figure 4). Further evidence in support of this is the fact that nicotine-induced catecholamine secretion occurs in Na^+-free, isotonic sucrose medium containing 5 mM Hepes buffer and 2.2 mM $CaCl_2$, while veratridine-induced secretion does not occur upon complete replacement of Na^+ with isosmolal sucrose or choline. Li^+ can replace Na^+, but only 50% of the maximal secretory response is obtained.[14] Although 56 mM K^+ depolarizes adrenal chromaffin cells and elicits Ca^{2+}-dependent catecholamine secretion, it does not promote $^{22}Na^+$ uptake.[12]

In addition to their effects on Na^+ channels BTX, veratridine, and aconitine are potent

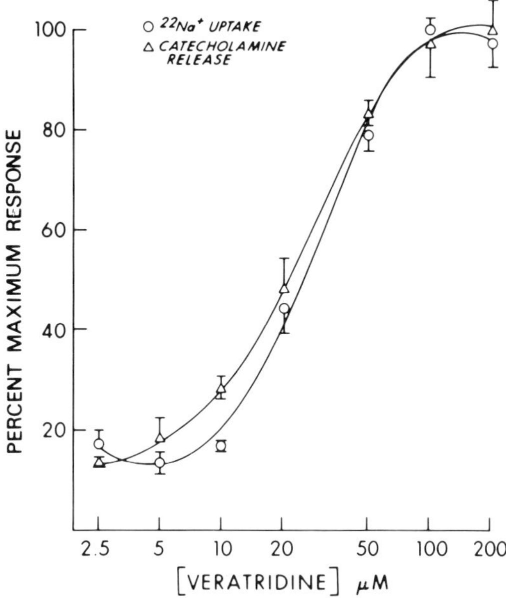

FIGURE 1. $^{22}Na^+$ uptake (O) and catecholamine secretion (△) after 10-min exposure to different concentrations of veratridine. Values are presented as the percent of the maximum veratridine-stimulated $^{22}Na^+$ uptake and catecholamine secretion. Maximum uptake was 9.9 ± 0.2 nmol of $^{22}Na^+$ per well; maximum catecholamine secretion was 23.8 ± 1.4% of the total catecholamine content per well. Control values were subtracted from the data which are representative of two separate experiments. Error bars are the SE of triplicate determinations. Similar results were obtained when the experiment was carried out in Locke's solution containing 154 mM NaCl. (From Amy, C. and Kirshner, N., *J. Neurochem.*, 39, 132, 1982. With permission.)

anticholinergic agents.[15] In Locke solution containing 1 μM TTX or in Na^+-free, Ca^{2+}-sucrose media alone, conditions under which the neurotoxins do not stimulate catecholamine secretion or promote $^{22}Na^+$ uptake, they block both nicotine-induced catecholamine secretion and $^{22}Na^+$ uptake.[12] The inhibition of catecholamine secretion was not competitive with respect to nicotine and was not readily reversed by washing. The potency order for inhibition of nicotine-induced catecholamine secretion was aconitine > BTX > veratridine with IC_{50} values of 3.0, 7.0, and 25 μM, respectively. The order differs from that for neurotoxin stimulation of catecholamine secretion, which was BTX > aconitine > veratridine with IC_{50} values of 0.7, 2.8, and 50 μM, respectively. In addition, aconitine is only a partial agonist for secretion eliciting only about 30% of the maximal response produced by veratridine or BTX, but is more efficacious than veratridine or BTX in inhibiting nicotine-induced secretion. In contrast to the inhibitory effects of aconitine, veratridine, and BTX, scorpion venom does not inhibit nicotine-induced catecholamine secretion. These studies suggest that aconitine, veratridine, and BTX block the nicotinic receptor ion conductance channel or inactivate the receptor by an allosteric modification. Veratridine and BTX have similar inhibitory effects at the frog neuromuscular junction. Garrison et al.[16] have reported that veratridine and BTX in the presence of TTX inhibit the depolarizing effect of bath-applied cholinergic agonists on frog muscle endplates.

Methoxyverapamil (D600) inhibits veratridine-activated fast sodium channels in cultured heart cells and cultured neuroblastoma cells.[17] This inhibition is competitive with veratridine

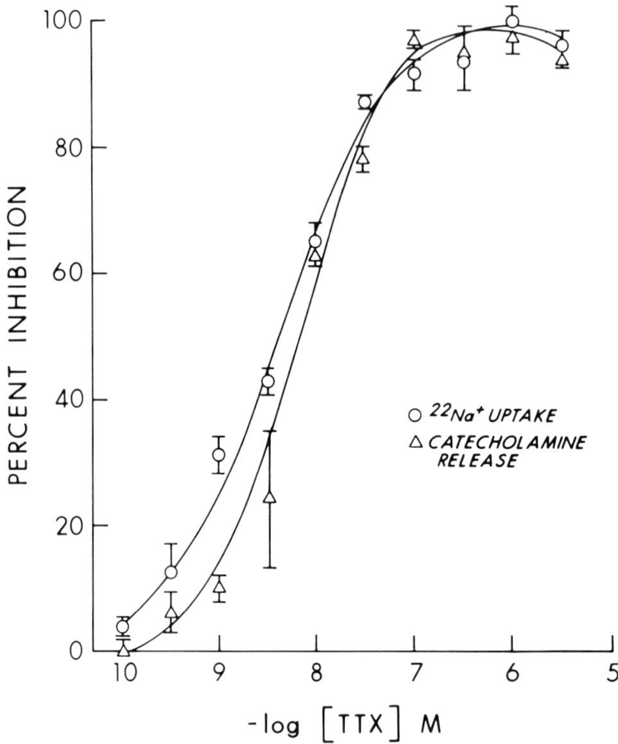

FIGURE 2.   Inhibition of veratridine-stimulated catecholaine secretion
($\triangle$) and ^{22}Na$^+$ uptake ($\bigcirc$) by TTX. Veratridine (100 μM) was added to
each well in the presence of the indicated amount of TTX. In the absence
of TTX, ^{22}Na$^+$ uptake was 15.6 $\pm$ 0.3 nmol of ^{22}Na$^+$/well/10 min; catecholamine secretion was 12.9 $\pm$ 0.2% of total content/well/10 min. Values
are the average $\pm$ the range of duplicate determinations. (From Amy, C.
and Kirshner, N., *J. Neurochem.*, 39, 132, 1982. With permission.)

and antagonized by Ca^{2+}. In adrenal chromaffin cells D600 inhibits both veratridine-induced
and nicotine-induced ^{22}Na$^+$ uptake, but by different mechanisms (Figure 5).[18] Similar to
neuroblastoma cells, D600 is a competitive inhibitor of Na$^+$ uptake with respect to veratridine, but unlike neuroblastoma cells this inhibition is not antagnoized by Ca^{2+}. D600
inhibition of nicotine-stimulated Na$^+$ uptake is not competitive with respect to nicotine nor
is it antagonized by Ca^{2+}.

The pharmacology of the Na$^+$ channel in adrenal medullary cells is similar to that of
TTX-sensitive Na$^+$ channels in other tissues. The concentration of TTX for half-maximal
inhibition of ^{22}Na$^+$ uptake in adrenal chromaffin cells[12] is within the range of 1 to 5 nM
found for inhibition of ^{22}Na$^+$ influx in nerve preparations[9] and in other types of cultured
cells.[19,20] The EC$_{50}$ secretion values for BTX, aconitine, veratridine, and scorpion venom
(0.9 μM, 2.8 μM, 50 μM, and 0.4 μg/mℓ, respectively) are within the range of values
reported for their depolarizing effects on nerve and muscle.[21-24] Catterall[13,25] has reported
half-maximal concentrations of 0.2 to 0.4 μM BTX, 8 μM aconitine, and 1 μg/mℓ scorpion
venom for the stimulation of ^{22}Na$^+$ uptake into neuroblastoma cells. Similar values were
also found for Na$^+$ channel activation in cultured muscle cells.[26] Aconitine is a weak Na$^+$
channel activator in neuroblastoma cells[25] and is a poor secretagogue in cultured bovine
adrenal cells.[14] However, scorpion venom from *L. quinquestriatus* is a potent secretagogue
in the adrenal cell culture system,[14] but only a weak Na$^+$ channel activator in neuroblastoma
cells.[25]

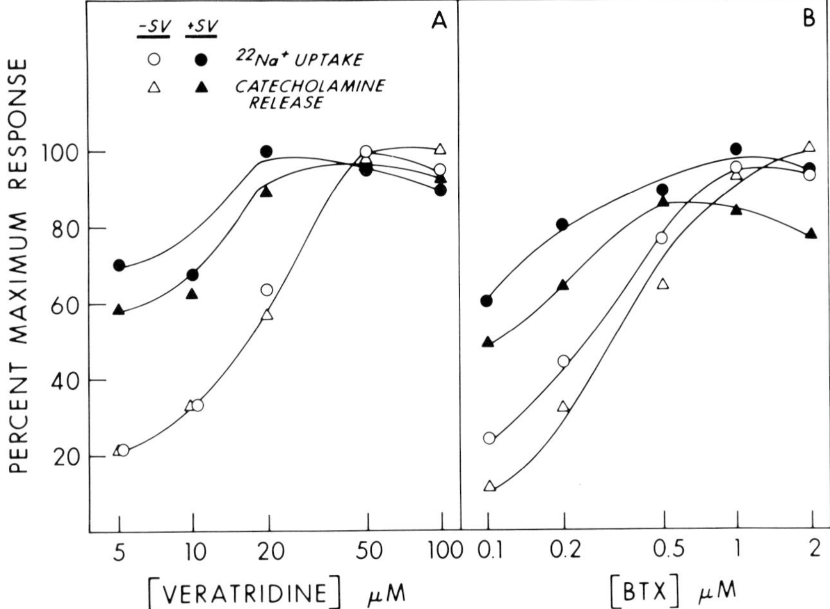

FIGURE 3.   Effects of scorpion venom on $^{22}Na^+$ uptake and catecholamine secretion in response to veratridine and BTX. Cells exposed to a range of concentrations of veratridine (A) and BTX (B) were assayed for $^{22}Na^+$ uptake (O●) and catecholamine release (△,▲) in the presence (closed symbols) or absence (open symbols) of 0.5 μg/mℓ of scorpion venom. Maximum catecholamine secretion due to veratridine (100 μM) was 46.8 ± 1.3%; BTX (2.0 μM) caused a maximum secretion of 42.3 ± 1.6%/well/10 min. Maximum $^{22}Na^+$ uptake was 12.4 ± 0.1 nmol/well/10 min with veratridine and 11.7 ± 0.4 nmol/well/10 min with BTX. Responses to cells treated with 0.5 μg/ mℓ scorpion venom alone were subtracted from the responses for cells treated with the alkaloid toxins and scorpion venom. (From Amy, C. and Kirshner, N., *J. Neurochem.*, 39, 132, 1982. With permission.)

B. Electrophysiological Analysis

The electrophysiological properties of adrenal medullary cells have been examined by a number of investigators. Douglas and co-workers,[27,28] using cultures of gerbil adrenal medullary cells, measured resting mean potentials of $-29.3 ± 0.2$ to $-32.8 ± 0.2$ mV, depending upon the density of the cell cultures. The membrane was depolarized upon addition of a variety of secretagogues, including acetylcholine, nicotine, pilocarpine, histamine, serotonin, angiotensin, bradykinin, Ba^{2+}, and a concentration of potassium exceeding 5.6 mM. Depolarization in response to acetylcholine fell linearly with the log of the extracellular sodium concentration, while depolarization by potassium was linearly related to the log of the potassium concentration. Using isolated, perfused rat adrenal glands, Ishikawa and Kanno[29] obtained an average resting potential of $-49.6 ± 1.9$ mV and found a linear relationship between the membrane potential and the log of the external K^+ concentration. A small depolarization also occurred in Ca^{2+}-containing Na^+-free sucrose medium upon addition of acetylcholine. It was concluded that depolarization in response to acetylcholine involves inward movement of both Na^+ and Ca^{2+} ions.[28] Pharmacological studies indicated that activation of either nicotinic or muscarinic receptors in gerbil adrenal medullary cells could result in depolarization. The depolarizing effect of acetylcholine was partially inhibited by hexamethonium and atropine. Hexamethonium alone completely blocked the response to nicotine, and atropine alone abolished the response to pilocarpine.[28]

Resting membrane potentials of -50 to -70 mV have been reported by others for cultures of rat and gerbil chromaffin cells, for cultured human pheochromocytoma cells, and for

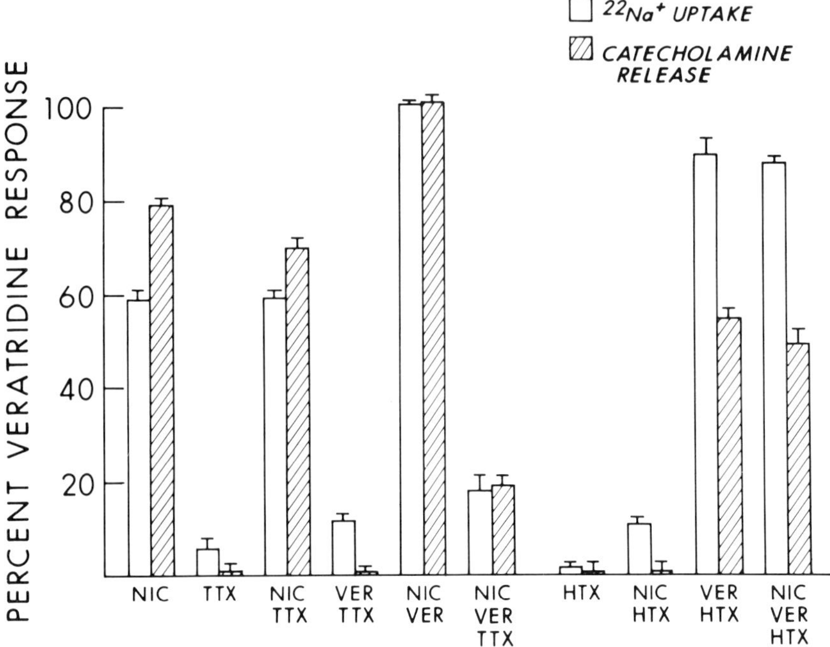

FIGURE 4.   The effects of TTX and HTX on nicotine- and veratridine-stimulated catecholamine secretion and ^{22}Na$^+$ uptake. Agonists and antagonists were present during the assay at the following concentrations: nicotine, 10 μM; TTX, 2 μM; veratridine, 100 μM; and HTX, 10 μM. Results are presented as the percent of the veratridine-stimulated response that equaled 11.7 $\pm$ 0.1 nmol of ^{22}Na$^+$ taken up/well and 43.7 $\pm$ 2.5% of the total catecholamines released/well over a 10-min period. Each bar represents the mean of the responses from three separate determinations $\pm$ SE. (From Amy, C. and Kirshner, N., *J. Neurochem.*, 39, 132, 1982. With permission.)

chromaffin cells in perfused intact rat adrenal glands. Brandt et al.,[30] Biales et al.,[31] and Kidokoro et al.[32] have shown that action potentials in chromaffin cells are evoked by injection of current into the cell and by application of acetylcholine or depolarizing concentrations of K$^+$. The major fraction of the action potential was absent upon replacement of Na$^+$ with Tris or upon addition of TTX to the medium. When Ca^{2+} in Na$^+$-free medium was replaced by Ba^{2+}, prolonged action potentials occurred. It was concluded that the action potentials are due mainly to an increase in membrane permeability to Na$^+$ and that they also have a Ca^{2+} component.

More recently, Na$^+$ channels have been studied in cultured bovine chromaffin cells using the patch clamp technique.[33,34] Resting membrane potentials of -50 to -80 mV were recorded. Several types of ionic channels were consistently found, including voltage-sensitive Na$^+$ and Ca$^+$ channels, acetylcholine-sensitive channels, and several types of voltage-sensitive K$^+$ channels. The results of these studies show that Na$^+$ channels in bovine chromaffin cells are very similar to those in nerve and muscle cell preparations. Single-channel Na$^+$ currents evoked by electrical depolarization are completely suppressed by TTX. The main Na$^+$ channel open time was 1 msec at -30 mV. Acetylcholine channels had a duration of 27 msec with 5 μM acetylcholine in the pipette and occurred in a burst-like fashion at higher concentrations. The effect of TTX on the acetylcholine channels was not reported. α-Bungarotoxin up to a concentration of 50 μg/mℓ did not suppress acetylcholine-induced currents, consistent with previous reports that this toxin did not inhibit nicotine- or acetylcholine-induced catecholamine secretion.[6,14,20,35]

Kidokoro and Ritchie[36] have examined the relationship of action potentials generated in

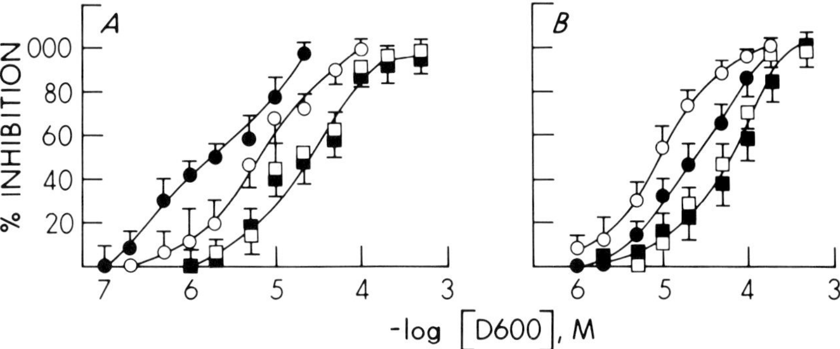

FIGURE 5.   Inhibition of Ca^{2+} uptake, Na^+ uptake, and catecholamine secretion by D600. Cells were incubated with 10 μM nicotine or 100 μM veratridine for 10 min in the presence of the indicated concentration of D600. Values are the means $\pm$ SE for triplicate analyses on three different cell preparations. (A) Nicotine-stimulated Ca^{2+} uptake (●) and catehcolamine secretion (○); veratridine-stimulated Ca^{2+} uptake (■) and catecholamine secretion (□). Absolute values in the absence of D600 were 0.73 $\pm$ 0.08 and 0.96 $\pm$ 0.07 nmol Ca^{2+} taken up upon stimulation by nicotine and veratridine, respectively, and 30.6 $\pm$ 2.8 and 27.0 $\pm$ 2.4% of total catecholamine content released; (B) nicotine-stimulated Na^+ uptake (●) and catecholaine secretion (○); veratridine-stimulated Na^+ uptake (■) and catecholamine secretion (□). Absolute values in the absence of D600 were 22.4 $\pm$ 2.8 and 33.6 $\pm$ 3.7 nmol Na^+ taken up/10 min upon stimulation by nicotine and veratridine and 24.5 $\pm$ 2.3 and 30.3 $\pm$ 1.8% of total catecholamine content released, respectively. (From Corcoran, J. J. and Kirshner, N., *J. Neurochem.*, 40, 1106, 1983. With permission.)

cultured rat chromaffin cells by application of acetylcholine and depolarizing concentrations of KCl to catecholamine secretion in perfused adrenal glands induced by these same agents. TTX (6 μM) markedly attenuated action potentials, but only partially inhibited secretion evoked by concentrations of KCl between 10 and 20 mM. At KCl concentrations of 30 mM or greater TTX did not affect secretion. Similarly, TTX only partially inhibited acetylcholine-induced secretion when the concentration of acetylcholine was 10 μM or less, but had no effect on secretion at higher concentrations of acetylcholine (50 μM). In contrast, Trifaró and Lee[6] found that 5 μM TTX inhibited secretion induced by 100 μM acetylcholine by 47%, but did not affect secretion induced by 56 mM KCl. Kidokoro and Ritchie[36] have proposed that modulation of spike frequency by KCl and acetylcholine may contribute to stimulation of catecholamine secretion, but since the effects of TTX were partial and confined to low levels of stimulation, the Na^+ action potential plays a facilitory and not an obligatory role in stimulation of secretion.

The results of the pharmacological and electrophysiological studies show that depending upon the mode of stimulation, Na^+ can gain entry into adrenal medullary chromaffin cells through two different types of channels, voltage-sensitive Na^+ channels which are similar to those present in other types of excitable tissue and the ion channel associated with the nicotinic receptor.[32] These studies suggest that upon stimulation with acetylcholine or other nicotinic agonists, Na^+ enters the cell initially through the nicotinic receptor channel. Entry of Na^+ depolarizes the cell and activates voltage-sensitive Na^+ channels, resulting in TTX-sensitive action potentials which may play a facilitory role in stimulation of secretion. Depolarization of the cell may also result in activation of voltage-sensitive Ca^{2+} channels, but it is not clear (discussed later), whether the entry of Ca^{2+} is gated through voltage-sensitive Ca^{2+} channels, the nicotinic receptor ion channel, or both. Upon stimulation with veratridine or batrachotoxin, entry of Na^+ is exclusively through TTX-sensitive Na^+ channels and Ca^{2+} entry occurs through voltage-sensitive Ca^{2+} channels.

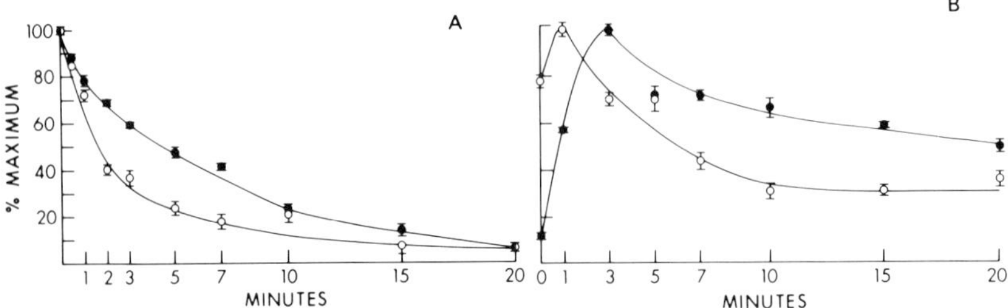

FIGURE 6.   (A) Rate of nicotine-stimulated $^{45}Ca^{2+}$ uptake ($\bigcirc$) and endogenous catecholamine release ($\bullet$) as a function of time. Cells were incubated in 0.3 mℓ of unlabeled medium containing 10 μM nicotine. At the indicated times after addition of nicotine, 30 $\mu\ell$ of medium containing $^{45}Ca^{2+}$ was added and the amount of $^{45}Ca^{2+}$ taken up during the following 1-min period, as well as the amounts of endogenous catecholamines released, were then determined. Maximal rates of $^{45}Ca^{2+}$ uptake and catecholmaine release are 2.1 nmol/min and 4.6% release of total content/min/10^6 cells, respectively. Nonstimulated controls have been subtracted; (B) rates of $^{45}Ca^{2+}$ uptake ($\bigcirc$) and catecholamine release ($\bullet$) stimulated by veratridine (100 μM). The experiment was carried out as described in (A). Maximum rates of $^{45}Ca^{2+}$ uptake and catecholamine release are 0.77 nmol/min/ 10^6 cells and 1.2% release of total content/min/10^6 cells, respectively. (From Kilpatrick, D. L., Slepetis, R. J., Corcoran, J. J., and Kirshner, N., *J. Neurochem.*, 38, 427, 1982. With permission.)

III. CALCIUM CHANNELS

A. Biochemical and Pharmacological Studies

The central role of Ca^{2+} in agonist-induced catecholamine secretion by the adrenal medulla was first demonstrated in 1961,[37] and many studies since have established that entry of Ca^{2+} into the cell is an early and obligatory event in stimulus-secretion coupling.[38,39] More recent studies employing adrenal medulla cell cultures have established that in competent cells elevation of intracellular Ca^{2+} is sufficient in itself to cause catecholamine secretion.[40-43] This section examines recent studies on the entry of Ca^{2+} into adrenal medullary cells following different modes of stimulation.

The requirement of external Ca^{2+} for catecholamine secretion by cultured adrenal medullary cells was established by several investigators during early studies of these cultures.[3-5] However, there have been but a handful of studies which have directly investigated in detail the relationship between Ca^{2+} uptake and catecholamine secretion. Several studies suggest that Ca^{2+} entry can occur either through voltage-sensitive Ca^{2+} channels or through the ion channel associated with the nicotinic cholinergic receptor, depending upon the mode of stimulation.

Upon stimulation with nicotinic agonists, veratridine, or 56 mM K^+, there is a rapid influx of Ca^{2+} and onset of catecholamine secretion,[44-46] but, depending on the mode of stimulation, there are differences in the time courses of these two events (Figure 6).[44] With nicotine, the maximal rates of Ca^{2+} uptake and catecholamine secretion occur within the first minute of stimulation. These rates gradually decline during the subsequent 15 min to levels of 10 to 15% of the maximum observed at 1 min. With veratridine the maximal rate of Ca^{2+} uptake occurs between the first and second minute of stimulation, while the maximal rate of catecholamine secretion occurs between the second and third minute of stimulation. With veratridine there is also a more prolonged decline in the rates of Ca^{2+} uptake and catecholamine secretion. After 15 min exposure to veratridine the rate of Ca^{2+} uptake had declined to 30% of the maximal rate, while the rate of catecholamine secretion had declined to 60% of the maximal rate. Because nicotine-induced Ca^{2+} uptake and catecholamine secretion were both maximal at the earliest time point used, one could not determine whether Ca^{2+} uptake preceded secretion, but with veratridine as agonist, Ca^{2+} uptake clearly preceded

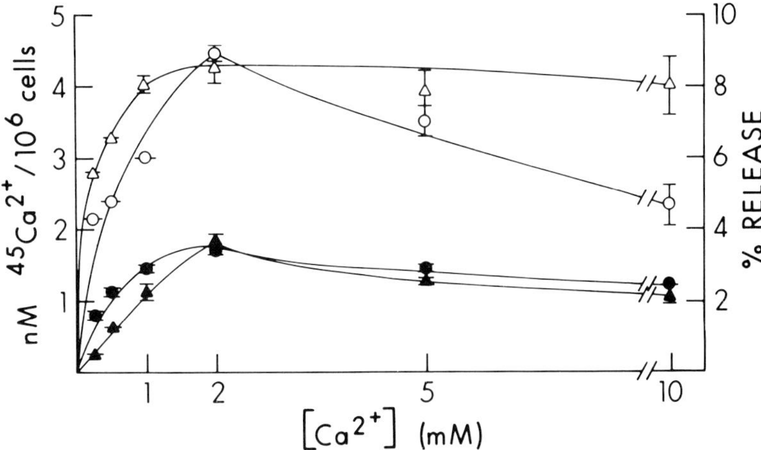

FIGURE 7.   Stimulation of $^{45}Ca^{2+}$ uptake (●, ▲) and endogenous catecholamine release (○, △) by veratridine, 100 μ*M* (triangles), and K$^+$, 50 m*M* (circles) at different calcium concentrations. Incubations were carried out for 5 min. (From Kilpatrick, D. L., Slepetis, R. J., Corcoran, J. J., and Kirshner, N., *J. Neurochem.*, 38, 427, 1982. With permission.)

catecholamine secretion. The maximal rate of $^{45}Ca^{2+}$ uptake upon stimulation with nicotine was significantly greater than that with veratridine, 2.1 and 0.77 nmol/min/10^6 cells, respectively. The slower influx of Ca^{2+} upon stimulation with veratridine may account for the delay in catecholamine secretion and also suggests that different channels are utilized for Ca^{2+} entry. Since veratridine promotes the same or greater rate of Na^+ entry as does nicotine during the first minute of stimulation,[12] the greater rate of $^{45}Ca^{2+}$ influx induced by nicotine would not be explained by activation of voltage-sensitive Ca^{2+} channels as a result of Na^+ entry.

There are also differences in the dose-response relationships for Ca^{2+} uptake and catecholamine secretion with different agonists.[44,45] With veratridine there is a close parallel relationship between Ca^{2+} uptake and catecholamine secretion, the EC_{50} values being the same. With nicotine as agonist there is also a parallel relationship between catecholamine secretion and Ca^{2+} uptake, but the dose-response curve for Ca^{2+} uptake is displaced to the right of that for catecholamine secretion, the EC_{50} values being 10 and 3 μ*M*, respectively. The dose-response curves for K$^+$-stimulated Ca^{2+} uptake and catecholamine secretion also parallel each other up to 45 m*M* K$^+$. At this concentration secretion was maximal, but uptake of Ca^{2+} continued to increase with increasing concentrations of K$^+$ up to 90 m*M*, the highest concentration tested. Ca^{2+} uptake at 45 m*M* K$^+$ was 60% of that at 90 m*M* K$^+$.

The dependency of Ca^{2+} uptake upon external Ca^{2+} concentration also varied with the mode of stimulation.[44,45] With veratridine and 56 m*M* K$^+$, both Ca^{2+} uptake and catecholamine secretion reached a plateau at 2 m*M* Ca^{2+} and remained constant with increasing extracellular Ca^{2+} concentrations up to 10 m*M* (Figure 7[44]). With nicotine and carbachol as agonists the uptake of Ca^{2+} increased continuously with increasing concentrations of Ca^{2+} up to 10 to 16 m*M* (Figure 8).[44,45] The Ca^{2+} concentration dependence for Ca^{2+} uptake was very similar in Locke solution and in Na^+-free sucrose medium upon stimulation with nicotine.[44]

Using isolated, perfused rat adrenals, Ishikawa and Kanno[29] found that after long exposure to acetylcholine when the secretory response had declined below 50% of the maximal, addition of 28 m*M* K$^+$ to the perfusion medium resulted in increased secretion. When the order of agonist treatments was reversed, acetylcholine increased catecholamine secretion in the presence of 28 m*M* K$^+$ after the initial response to K$^+$ had declined. It was concluded

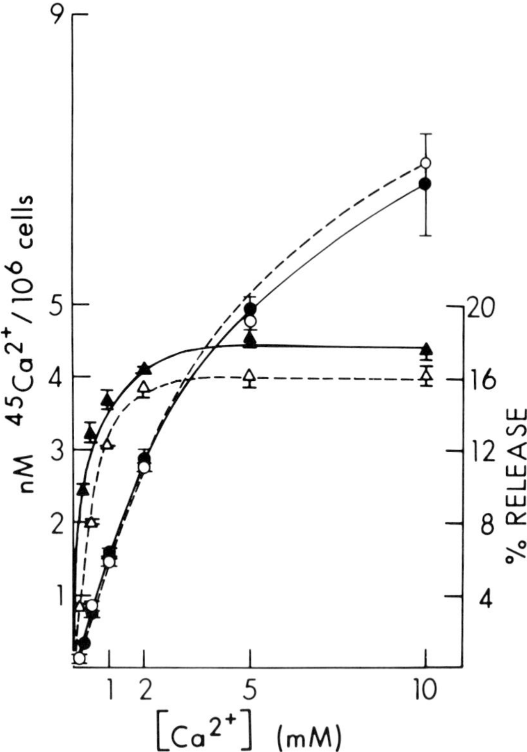

FIGURE 8.   Nicotine-stimulated uptake of $^{45}Ca^{2+}$ (circles) and release of catecholamines (triangles) as a function of calcium concentration. Cells were stimulated for 5 min with 10 μM nicotine in Locke's solution (●, ▲) or sucrose medium (○, △). (From Kilpatrick, D. L., Slepetis, R. J., Corcoran, J. J., and Kirshner, H., *J. Neurochem.*, 38, 427, 1982. With permission.)

that the mechanism of Ca^{2+} influx activated by excess K^+ may differ from that activated by acetylcholine.

The effects of pharmacological agents on Ca^{2+} uptake were similar to their effects on Na^+ uptake.[12,44] Histrionicotoxin blocked nicotine-induced Ca^{2+} uptake, but did not affect veratridine- and K^+-induced uptake. TTX blocked veratridine-induced Ca^{2+} uptake, but did not block nicotine- or K^+-stimulated uptake. D600 inhibited both nicotine- and veratridine-stimulated Ca^{2+} uptake, but nicotine-stimulated Ca^{2+} uptake was about tenfold more sensitive to inhibition by D600 than was veratridine-stimulated uptake (Figure 5).[18] As with Na^+ uptake, the inhibition of Ca^{2+} uptake by D600 was competitive with respect to veratridine and not competitive with respect to nicotine. With each secretagogue, Na^+ uptake was less sensitive to inhibition than was Ca^{2+} uptake, the ratios IC_{50} Na^+ uptake/IC_{50} Ca^{2+} uptake being 7.2 and 2.9, respectively, for nicotine- and veratridine-induced stimulation. However, since Na^+ uptake was determined in cells treated with ouabain, while Ca^{2+} uptake was determined in untreated cells, one cannot conclude from these studies that Ca^{2+} uptake is more sensitive to inhibition by D600 than is Na^+ uptake.

The sum of these studies suggests that treatment of cells with depolarizing agents activates, as a result of Na^+ entry, voltage-sensitive calcium channels similar to the slow Ca^{2+} channels found in invertebrate axons, while nicotinic agonists promote entry of Ca^{2+} through the ion channel associated with the nicotinic cholinergic receptor. Uptake of Ca^{2+} stimulated by K^+ or veratridine saturated at external Ca^{2+} concentrations of 2 mM, while Ca^{2+} uptake stimulated by nicotine did not saturate at external Ca^{2+} concentrations of 10 to 16 mM Ca^{2+}. Furthermore, nicotine-stimulated Ca^{2+}-uptake occurred in sucrose media and was essentially similar to nicotine-stimulated Ca^{2+} uptake in Locke medium, both with respect to the Ca^{2+} concentration dependence for uptake (Figure 8) and with respect to inhibition by D600. The greater sensitivity of nicotine-stimulated Ca^{2+} uptake to inhibition by D600 than that of veratridine-stimulated uptake may have alternative explanations: (1) entry of calcium through the nicotinic receptor channel is inhibited more readily than is entry through voltage-sensitive channels. This would support independent channels for Ca^{2+} entry; (2) D600 is a noncompetitive nicotinic antagonist. Inhibition of receptor activation would inhibit subsequent events which lead to activation of voltage-sensitive channels.

Calcium influx through acetylcholine receptor channels may directly stimulate secretion, or in combination with Na^+ influx through the same channel may depolarize the cell to activate voltage-sensitive Ca^{2+} channels. In Locke solution, entry of Na^+ and Ca^{2+} through the receptor channel may be sufficient to activate voltage-sensitive Ca^{2+} channels, whereas in sucrose medium sufficient depolarization may occur as a result of inward flux of Ca^{2+} and outward flux of K^+. Different channels for Ca^{2+} entry, depending upon the mode of stimulation, have also been proposed for PC12 cells.[20]

Calcium influx has also been determined using the fluorescent dye Quin 2.[47] The resting internal Ca^{2+} concentration ($[Ca^{2+}]_i$) was close to 100 nM. Exposure of cells to acetylcholine or 80 mM K^+ resulted in a transient increase of $[Ca^{2+}]_i$, which reached a maximum within 6 sec and declined with a half-life of 2 to 5 min. Hexamethionium, D600, or the absence of extracellular Ca^{2+} prevented both the rise in $[Ca^{2+}]_i$ and the secretion of catecholamines. From an analysis of the relationship between the intracellular concentration of Quin 2 and the apparent amount of Ca^{2+} entering the cells, the authors proposed that two processes may be involved in the entry of Ca^{2+}: one that is activated by depolarization and inhibited by a rise in $[Ca^{2+}]_i$ and a second that is activated strongly by the rise in $[Ca^{2+}]_i$.

A number of divalent and trivalent cations as well as a variety of organic agents inhibit Ca^{2+} uptake and catecholamine secretion. Mg^{2+}, Co^{2+}, Ni^+, Cd^{2+}, Mn^{2+}, Sr^{2+}, and Gd^{3+} have been reported to inhibit these processes. Gd^{3+} is especially potent, inhibiting uptake by 90% at 0.05 mM.[48] In one study[49] the order of potency for inhibition was $Cd^{2+} > Mn^{2+} > Co^{2+} > Sr^{2+}$, which is similar to that seen in brain synaptosomes.[50] Sr^{2+} and Mn^{2+} can also substitute for Ca^{2+} in evoking catecholamine secretion.[49] The time course and extent of nicotine-induced secretion in the presence of Sr^{2+} was similar to that in the presence of Ca^{2+}, but was markedly different from that of Mn^{2+}. The onset of secretion in the presence of Mn^{2+} had a lag period of 3 to 4 min, but secretion continued over a period of 1 hr, while with Ca^{2+}, onset of secretion was immediate and was essentially complete in 10 min. Uptake of Mn^{2+} also occurred over a period of 1 hr compared to 5 to 10 min for Ca^{2+}, but both Mn^{2+} and Ca^{2+} uptake were inhibited to a similar extent by equimolar concentrations of Ca^{2+}.

The effects of calmodulin antagonists on calcium uptake and catecholamine secretion have been reported by several investigators, but the interpretation of the results are controversial. Trifluoperazine (TFP), other phenothiazines, and pimozide caused a parallel inhibition of Ca^{2+} uptake and catecholamine secretion in response to nicotinic agonists, veratridine, and 56 mM K^+ (Figure 9).[51,52] The responses to carbamylcholine and nicotine were 3- to 20-fold more sensitive to inhibition by the various agents than was the response to K^+. The calmodulin inhibitors also inhibited secretion induced by the calcium ionophore Ionomycin, but only at high concentrations (40 to 50 μM) similar to those required for inhibition of K^+ responses. It was concluded from these studies that the calmodulin inhibitors have multiple effects on chromaffin cells. At low concentrations they inhibit Ca^{2+} transport induced by veratridine, carbamylcholine, and nicotine by interfering with membrane events initiated by these agonists. This inhibition was noncompetitive with respect to nicotine and veratridine.

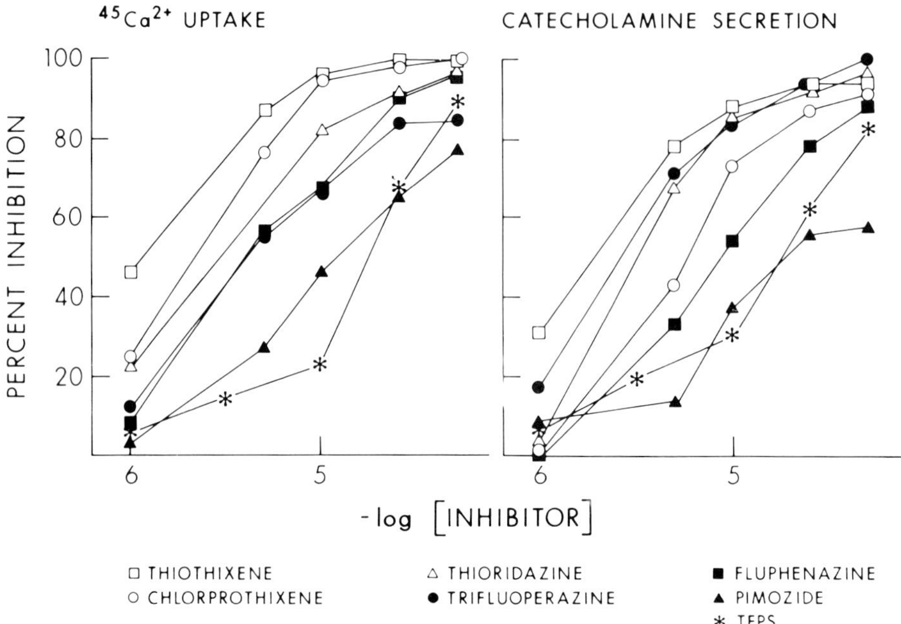

FIGURE 9.   Inhibition of nicotine-stimulated $^{45}Ca^{2+}$ uptake and catecholamine secretion by phenothiazine sand pimozide. Cells were incubated for 5 min with the indicated concentrations of drugs before addition of 10 μM nicotine. Secretion was terminated after 10 min at ambient room temperature. In the absence of inhibitors 10 μM nicotine released 20 to 30% of the total cellular content of catecholamines after subtraction of the release from unstimulated controls. (From Slepetis, R. and Kirshner, N., *Cell Calcium*, 3, 183, 1982. With permission.)

Consistent with the effects on nicotine-induced secretion are reports that chlorpromazine acts as a noncompetitive inhibitor of the acetylcholine current and binds to a high affinity site on the nicotinic receptor in the presence of agonists.[53] At higher concentrations TFP and other phenothiazines inhibit Ca^{2+} entry and secretion initiated by K^+, Ba^{2+}, and ionomycin by nonspecific membrane perturbations or possibly by a calmodulin-dependent reaction. In cells permeabilized by high-intensity electrical discharges, 10 μM TFP inhibited secretion by 40%, but 10 μM pimozide, which binds more tightly to calmodulin, did not.[41] However, in cells permeabilized with digitonin neither 10 μM TFP nor 10 μM pimozide inhibited Ca^{2+}-induced secretion.[42]

Kenigsberg et al.[54] have reported that TFP, at concentrations which block K^+-induced catecholamine secretion, did not affect K^+-induced Ca^{2+} uptake or Ca^{2+} efflux. They concluded that TFP blocks secretion distal to the entry of Ca^{2+} and suggested that calmodulin has a role in the secretory process. The concentrations of TFP reported by Kenigsberg et al.[54] for half-maximal inhibition of secretion were tenfold lower than those reported by Wada et al.[52] and by Slepetis and Kirshner,[51] and a different procedure was used to evaluate Ca^{2+} uptake. Kenigsberg et al.[54] evaluated Ca^{2+} uptake by a washout procedure over a period of 65 min after stimulation, while Wada et al.[52] and Slepetis and Kirshner[51] directly determined $^{45}Ca^{2+}$ uptake immediately after cessation of stimulation. Since there is a rapid loss of the induced Ca^{2+} uptake following cessation of stimulation,[44] the washout procedure would not have detected the fraction of Ca^{2+} taken up as a result of stimulation.

Burgoyne and Norman[55] have reported that the calmodulin inhibitor calmidazolium, at a concentration of 1 μM, inhibited carbamylcholine-induced catecholamine secretion without affecting Ca^{2+} uptake determined by fluorescence with Quin 2. Calmidazolium was less effective in inhibiting secretion induced by A23187 and by 55 mM K^+, but measurements of Ca^{2+} influx induced by these agents were not reported. In our studies with calmidazolium (unreported) we found that higher concentrations of calmidazolium ($IC_{50} = 3$ μM) were required to inhibit nicotine-induced catecholamine secretion and that there was a parallel and equipotent inhibition of $^{45}Ca^{2+}$ uptake at all concentrations of calmidazolium tested up to 10 μM. Similar to the results of Burgoyne and Norman, we also found that higher concentrations of calmidazolium were required to inhibit secretion induced by 56 mM K^+ than were required for inhibition of nicotine-induced secretion, but here, too, there was also a parallel and equipotent inhibition of $^{45}Ca^{2+}$ uptake. Our studies do not support the conclusion by Burgoyne and Norman[55] that calmidazolium inhibits secretion at a point subsequent to the entry of Ca^{2+}.

Evidence that calmodulin may play a role in stimulus-secretion coupling also comes from a recent study by Kenigsberg and Trifaró.[56] They have shown that microinjection of calmodulin antibodies into cultured chromaffin cells caused significant inhibition of catecholamine release in response to stimulation by acetylcholine or 56 mM K^+. As noted by the authors these studies suggest that calmodulin-mediated reactions are involved in secretion of catecholamines, but they do not exclude the possibility that the antigen-antibody reaction might have inhibited secretion nonspecifically. Although microinjection of IgG from non-immunized animals was used as a control in these studies, additional controls using IgG directed against other cytoplasmic protein components, such as lactic dehydrogenase or tyrosine hydroxylase which are unlikely to be directly involved in secretion, need to be evaluated.

B. Electrophysiological Studies

Extensive analysis of Ca^{2+} channels in cultured bovine adrenal medulla chromaffin cells has been carried out by the patch-clamp technique.[33,34] In general, the properties of the Ca^{2+} channel are similar to those in other excitable tissues. Ca^{2+} currents resulting from voltage step depolarizing pulses were determined and provide important information on voltage-sensitive Ca^{2+} channels in these cells. However, the extent to which Ca^{2+} enters the cell through the nicotinic receptor channel was not addressed.

The effects of TFP on ionic current and secretion have also been studied by the patch-clamp technique.[57] In these studies increases in membrane capacitance which are expected to result from fusion of the chromaffin vesicle membrane with the plasma membrane were taken as indices of secretion. Although Ca^{2+}-dependent capacitance changes were observed upon electrical depolarization of the cell, no capacitance changes were detected upon treatment of the cells with 50 μM acetylcholine, a concentration which evokes near maximal catecholamine secretion. TFP at concentrations up to 10 μM had no effect on electrically induced Na^+ currents of the Hodgkin-Huxley type. However, TFP was a potent anticholinergic agent. At concentrations of 10 nM to 1 μM TFP decreased acetylcholine-induced currents. The inhibition was not immediate, but required several 10-sec applications of TFP plus acetylcholine before the effect was seen. Alternatively, preincubation of the cells for 30 sec with TFP resulted in immediate inhibition to a test dose of acetylcholine plus TFP. At concentrations of 10 μM, TFP decreased the calcium current by about 50% and almost completely prevented the capacitance changes resulting from electrical depolarization. The authors concluded that inhibition of secretion by TFP in intact cells should be due to inhibition of action potentials by its effects on acetylcholine and Ca^{2+} channels and by its intracellular prevention of a step or steps leading to vesicle fusion with the plasma membrane. However, it was also noted that these studies provide no evidence whether the effects of TFP are specifically mediated by calmodulin or by some other action of TFP.

The patch-clamp studies are consistent with the observations that concentrations of TFP in the submicromolar to micromolar range inhibit Ca^{2+} uptake and catecholamine secretion induced by nicotinic cholinergic agonists. Chlorpromazine is a noncompetitive inhibitor of the acetylcholine current and binds to several sites on the Torpedo receptor noncompetitively with carbamylcholine.[53] Blockade of the nicotinic receptor by TFP is in itself sufficient to explain its inhibitory effect on Ca^{2+} influx and catecholamine secretion induced by cholinergic agonists, and it appears superfluous to suggest that an additional effect on calmodulin is involved. At higher concentrations, TFP may block the Ca^{2+}-dependent fusion of the vesicles with the plasma membrane and prevent the capacitance changes induced by electrical depolarization. However, these studies do not exclude the possibility that the inhibitory effects of TFP on secretion induced by high concentrations of KCl or capacitance changes induced by electrical depolarization were not due to decreased Ca^{2+} influx. Although the Ca^{2+} current was not completely blocked at concentrations of TFP which blocked capacitance changes, the Ca^{2+} current may have been decreased to a level where it was insufficient to induce fusion. These studies indicate that TFP inhibits secretion by several mechanisms. At low concentrations it blocks secretion induced by nicotinic agonists through anticholinergic properties and at similar concentrations may also block the actions of depolarizing agents such as BTX and veratridine. The patch-clamp studies indicate that 10 μM TFP does not block the voltage-sensitive Na^+ channel, but it may, nevertheless, interfere with activation of the channel by depolarizing agents. At higher concentrations TFP may inhibit secretion by additional effects on vesicle translocation or by nonspecific membrane perturbations. Although one may speculate that the effect of TFP at higher concentrations is mediated through inhibition of calmodulin-dependent reactions, the evidence in support of this is equivocal.

IV. SUMMARY

Adrenal medullary chromaffin cell cultures provide an excellent system for studying the ionic events involved in stimulus-secretion coupling using both biochemical and electrophysiological techniques. Although much progress has been made, the role of the voltage-sensitive Na^+ channels in nicotinic receptor-induced secretion is not clear, nor is it clear how Ca^{2+} gains entry to the cell upon activation of the nicotinic receptor.

REFERENCES

1. **Brooks, J.,** The isolated bovine adrenal medullary chromaffin cell: a model of neuronal excitation-secretion, *Endocrinology,* 101, 1369, 1977.
2. **Waymire, J. C., Waymire, K. G., Boehme, R., Noritaki, D., and Wardell, J.,** Regulation of tyrosine hydroxylase by cyclic 3':5' adenosine monophosphate in cultured neuroblastoma and cultured dissociated bovine adrenal chromaffin cells, in *Structure and Function of Monoamine Enzymes,* Usdin, E., Ed., Marcel Dekker, New York, 1977.
3. **Schneider, A. S., Herz, R., and Rosenheck, K.,** Stimulus-secretion coupling in chromaffin cells isolated from bovine adrenal medulla, *Proc. Natl. Acad. Sci. U.S.A.,* 74, 5036, 1977.
4. **Fenwick, E. M., Fajdiga, P. B., Howe, N. B. S., and Livett, B. G.,** Functional and morphological characterization of isolated bovine adrenal medullary cells, *J. Cell Biol.,* 76, 12, 1978.
5. **Kilpatrick, D. L., Ledbetter, F. H., Carson, K. A., Kirshner, A. G., Slepetis, R., and Kirshner, N.,** Stability of bovine adrenal medulla cells in culture, *J. Neurochem.,* 35, 679, 1980.
6. **Trifaró, J. M. and Lee, R. W. H.,** Morphological characteristics and stimulus-secretion coupling in bovine adrenal chromaffin cell cultures, *Neuroscience,* 5, 1533, 1980.
7. **Wilson, S. P. and Viveros, O. H.,** Primary culture of adrenal medullary chromaffin cells in a chemically defined medium, *Exp. Cell Res.,* 133, 159, 1981.
8. **Wilson, S. P. and Kirshner, N.,** Preparation and maintenance of adrenal medullary chromaffin cell cultures, *Methods Enzymol.,* 103, 305, 1983.
9. **Catterall, W. A.,** Neurotoxins that act on voltage-sensitive sodium channels in excitable membranes, *Ann. Rev. Pharm. Toxicol.,* 20, 15, 1980.

10. **Albuquerque, E. X., Varnard, E. A., Chiau, T. H., Lapa, A. J., Doly, J. O., Jansson, S. R., Daly, J., and Witkop, B.,** Acetylcholine receptor and ion conductance modulator sites at the murine neuromuscular junction: evidence from specific toxic reactions, *Proc. Natl. Acad. Sci. U.S.A.,* 70, 949, 1973.
11. **Heidmann, T. and Changeux, J. P.,** Structural and functional properties of the acetylcholine receptor protein in its purified and membrane-bound states, *Ann. Rev. Biochem.,* 47, 317, 1978.
12. **Amy, C. and Kirshner, N.,** ^{22}Na$^+$ uptake and catecholamine secretion by primary cultures of adrenal medulla cells, *J. Neurochem.,* 39, 132, 1982.
13. **Catterall, W. A.,** Activation of the action potential sodium-ionophore of cultured neuroblastoma cells by veratridine and batrachotoxin, *J. Biol. Chem.,* 250, 4053, 1975.
14. **Kilpatrick, D. L., Slepetis, R., and Kirshner, N.,** Ion channels and membrane potential in stimulus-secretion coupling in adrenal medulla cells, *J. Neurochem.,* 36, 1245, 1981.
15. **Kilpatrick, D. L., Slepetis, R., and Kirshner, N.,** Inhibition of catecholamine secretion from adrenal medulla cells by neurotoxins and cholinergic antagonists, *J. Neurochem.,* 37, 125, 1981.
16. **Garrison, D. R., Albuquerque, E. X., Warnick, J. E., Daly, J. W., and Witcop, B.,** Antagonism of carbamylcholine-induced depolarization by batrachotoxin and veratridine, *Mol. Pharmacol.,* 14, 111, 1978.
17. **Galper, J. B. N. and Catterall, W. A.,** Inhibition of sodium channels by D600, *Mol. Pharmacol.,* 15, 174, 1979.
18. **Corcoran, J. J. and Kirshner, N.,** Inhibition of calcium intake, sodium uptake, and catecholamine secretion by methoxyverapamil (D600) in primary cultures of adrenal medulla cells, *J. Neurochem.,* 40, 1106, 1983.
19. **Catterall, W. A. and Nirenberg, M.,** Sodium uptake associated with activation of action potential ionophores of cultured neuroblastoma and muscle cells, *Proc. Natl. Acad. Sci. U.S.A.,* 70, 3759, 1973.
20. **Stallcup, W. B.,** Sodium and calcium fluxes in a clonal nerve cell line, *J. Physiol., (London),* 286, 525, 1979.
21. **Herzog, W. H., Feibel, R. M., and Bryant, S. H.,** The effect of aconitine on the giant axon of the squid, *J. Gen. Physiol.,* 47, 719, 1964.
22. **Ulbericht, W.,** The effect of veratridine on excitable membranes of nerve and muscles, *Ergeb. Physiol. Biol. Chem. Exp. Pharmakol.,* 61, 18, 1969.
23. **Ohta, M., Narahashi, T., and Keeler, R. F.,** Effect of veratrium alkaloids on membrane potential and conductance of squid and crayfish giant axons, *J. Pharmacol. Exp. Ther.,* 184, 143, 1973.
24. **Albuquerque, E. X. and Daly, J. W.,** Batrachotoxin, a selective probe for channels modulating sodium conduction in electrogenic membranes, in *Receptors and Recognition,* Series B, Chapman & Hall, London, 1977, 297.
25. **Catterall, W. A.,** Cooperative activation of action potential sodium ionophore by neurotoxins, *Proc. Natl. Acad. Sci. U.S.A.,* 72, 1782, 1975.
26. **Catterall, W. A.,** Purfication of a toxic protein from scorpion venom which activates the action potential sodium ionophore, *J. Biol. Chem.,* 251, 5528, 1976.
27. **Douglas, W. W., Kanno, T., and Sampson, S. R.,** Effect of acetylcholine and other medullary secretagogues and antagonists on the membrane potential of adrenal chromaffin cells: an analysis applying techniques of tissue culture, *J. Physiol. (London),* 188, 107, 1967.
28. **Douglas, W. W., Kanno, T., and Sampson, S. R.,** Influence of the ionic environment on the membrane potential of adrenal chromaffin cells and on the depolarizing effect of acetylcholine, *J. Physiol. (London),* 191, 107, 1967.
29. **Ishikawa, K. and Kanno, T.,** Influences of extracellular calcium and potassium concentrations on adrenaline release and membrane potentials in the perfused adrenal medulla of the rat., *Jpn. J. Physiol.,* 28, 275, 1978.
30. **Brandt, B. L., Hagiwara, S., Kodokoro, Y., and Miyazaki, S.,** Action potentials in the rat chromaffin cells and the effects of acetylcholine, *J. Physiol. (London),* 273, 417, 1976.
31. **Biales, B., Dichter, M., and Tischler, A.,** Electrical excitability of cultured adrenal chromaffin cells, *J. Physiol. (London),* 262, 743, 1976.
32. **Kidokoro, Y., Miyazaki, S., and Ozawa, S.,** Acetylcholine-induced membrane depolarization and potential fluctuations in the rat adrenal chromaffin cell, *J. Physiol. (London),* 324, 221, 1982.
33. **Fenwick, E. M., Marty, A., and Neher, E.,** A patch-clamp study of bovine chromaffin cells and of their sensitivity to acetylcholine, *J. Physiol. (London),* 331, 577, 1982.
34. **Fenwick, E. M., Marty, A., and Heher, E.,** Sodium and calcium channels in bovine chromaffin cells, *J. Physiol. (London),* 331, 599, 1982.
35. **Liang, B. T. and Perlman, R. L.,** Catecholamine secretion by hamster adrenal cells, *J. Neurochem.,* 32, 927, 1979.
36. **Kidokoro, Y. and Ritchie, A. K.,** Chromaffin cell action potentials and their possible role in adrenal secretion from rat adrenal medulla, *J. Physiol. (London),* 307, 199, 1980.
37. **Douglas, W. W. and Rubin, R. P.,** The role of calcium in the secretory response of the adrenal medulla to acetylcholine, *J. Physiol (London),* 159, 40, 1961.

38. **Douglas, W. W.,** Secretomotor control of adrenal medullary secretion: synaptic membrane and ionic events in stimulus-secretion coupling, in *Handbook of Physiology, Section 7: Endocrinology,* Vol. 6, Blaschko, H., Sayers, G., and Smith, A. D., Eds., American Physiological Society, Washington, D.C., 1975, 367.

39. **Viveros, O. H.,** Mechanism of secretion of catecholamines from adrenal medulla in *Handbook of Physiology, Section 7: Endocrinology,* Vol. 6, Blaschko, H., Sayers, G., and Smith, A. D., Eds., American Physiological Society, Washington, D.C., 1975, 389.

40. **Conn, P. M., Kilpatrick, D., and Kirshner, N.,** Ionophoretic calcium mobilization in rat gonadotropes and bovine adrenomedullary cells, *Cell Calcium,* 1, 129, 1980.

41. **Knight, D. E. and Baker, P. F.,** Calcium-dependence of catecholamine release from bovine adrenal medullary cells after exposure to intense electric fields, *J. Membr. Biol.,* 68, 107, 1982.

42. **Wilson, S. P. and Kirshner, N.,** Calcium-evoked secretion from digitonin-permeabilized adrenal medullary chromaffin cells, *J. Biol. Chem.,* 258, 4994, 1983.

43. **Dunn, L. A. and Holz, R. W.,** Catecholamine secretion from digitonin-treated adrenal medullary chromaffin cells, *J. Biol. Chem.,* 268, 4989, 1983.

44. **Kilpatrick, D. L., Slepetis, R. J., Corcoran, J. J., and Kirshner, N.,** Calcium uptake and catecholamine secretion by cultured bovine adrenal medulla cells, *J. Neurochem.,* 38, 427, 1982.

45. **Holz, R. W., Senter, R. A., and Fry, R. A.,** Relationship between calcium uptake and catecholamine secretion in primary dissociated cultures of adrenal medulla, *J. Neurochem.,* 39, 635, 1982.

46. **Izumi, F., Oka, M., and Kumakura, K.,** Calcium flux and catecholamine release in isolated bovine adrenal medullary cells: effects of nicotine and muscarinic stimulation, in *Advances in the Biosciences,* Vol. 36, Pergamon Press, New York, 1982, 29.

47. **Knight, D. E. and Kesteven, N. T.,** Evoked transient intracellular-free calcium changes and secretion in isolated bovine adrenal medullary cells, *Proc. R. Soc. London Ser. B,* 218, 177, 1983.

48. **Bourne, G. W. and Trifaró, J. M.,** The gadolinium ion: a potent blocker of calcium channels and catecholamine release from cultured chromaffin cells, *Neuroscience,* 7, 1615, 1982.

49. **Corcoran, J. J. and Kirshner, N.,** Effects of manganese and other divalent cations on calcium uptake and catecholamine secretion by primary cultures of bovine adrenal medulla cells, *Cell Calcium,* 4, 127, 1983.

50. **Blaustein, M. P.,** Effects of potassium, veratridine and scorpion toxin on calcium accumulation and transmitter release by nerve terminals in vitro, *J. Physiol. (London),* 247, 617, 1975.

51. **Slepetis, R. and Kirshner, N.,** Inhibition of $^{45}Ca^{2+}$ uptake and catecholamine secretion by phenothiazines and pimozide in adrenal medulla cells cultures, *Cell Calcium,* 3, 183, 1982.

52. **Wada, A., Yanagihara, N., Izumi, F., Sakurai, S., and Kobayashi, H.,** Trifluoperazine inhibits calcium uptake and catecholamine secretion and synthesis in adrenal medullary cells, *J. Neurochem.,* 40, 481, 1983.

53. **Heidmann, T., Oswald, R., and Changeux, J.-P.,** Multiple sites of action for noncompetitive blockers on acetylcholine receptor-rich membrane fragments from torpedo marmorata, *Biochemistry,* 22, 3112, 1983.

54. **Kenigsberg, R. L., Cote, A., and Trifaró, J. M.,** Trifluoperazine, a calmodulent inhibitor, blocks secretion in cultured chromaffin cells at a step distal from calcium entry, *Neuroscience,* 7, 2277, 1982.

55. **Burgoyne, R. D. and Norman, K.-M.,** Effect of calmidazolium and phorbol ester on catecholamine secretion from adrenal chromaffin cells, *Biochim. Biophys. Acta,* 805, 37, 1984.

56. **Kenigsberg, R. L. and Trifaró, J. M.,** Micro-injection of calmodulin antibodies into cultured chromaffin cells blocks catecholamine release in response to stimulation, *Neuroscience,* 14, 355, 1985.

57. **Clapham, D. E. and Neher, E.,** Trifluoperazine reduces inward ionic currents and secretion by separate mechanisms in bovine chromaffin cells, *J. Physiol. (London),* 353, 541, 1984.

58. **Burgoyne, R. D., Geisow, M. J., and Barron, J.,** Dissection of stages in exocytosis in the adrenal chromaffin cell with use of trifluoperazine, *Proc. R. Soc. London Ser. B,* 206, 111, 1982.

Chapter 13

RECENT ADVANCES IN MEMBRANE BIOPHYSICS OF THE ADRENAL CHROMAFFIN CELL

Yoshiaki Kidokoro

TABLE OF CONTENTS

I. INTRODUCTION

By virtue of the improved patch-clamp technique the various biophysical properties of adrenal chromaffin cells are under the grasp of our current investigation. Since I last reviewed the electrophysiology of chromaffin cells (1985),[20] more detailed analyses of membrane properties have been published. In this chapter I will mainly review these new materials.

It is well known in many secretory systems that Ca^{2+} plays an important role in coupling between stimulus and hormone secretion. Ca^{2+} must enter the cell to initiate secretion in the adrenal chromaffin cell. One of the goals in our field of electrophysiology is to elucidate the mechanism of Ca^{2+} entry. Since in the normal resting state the cell membrane is impermeable to Ca^{2+}, the stimulus has to change its ion permeability.

There are two major routes for Ca^{2+} to enter the adrenal chromaffin cell. One is voltage-dependent Ca channels and the other is acetylcholine (ACh) receptor channels. A strong suggestion that a voltage-dependent permeability increase to Ca^{2+} may be involved in hormone secretion stems from an observation that depolarization of the cell membrane with excess K^+ leads to hormone secretion and that external Ca^{2+} is required for this stimulation.[12] A voltage-dependent Ca^{2+} conductance increase was clearly demonstrated in the rat adrenal chromaffin cell in culture as a regenerative action potential in Na-free saline which was blocked by $CoCl_2$.[4,20] In agreement with this finding under voltage-clamp conditions, a slow inward current which is distinctly different from major Na inward current was observed.[15] Depolarization caused by external high K saline will activate these channels and permit Ca^{2+} entry. Likewise, during action potentials which are mainly due to Na^+ inward current a certain amount of Ca^{2+} will enter the cell through voltage-dependent Ca channels. Indeed, the natural stimulant, ACh, changes the action potential frequency in a dose-dependent manner. The concentration range of ACh to modify the action potential frequency is similar to that for adrenaline secretion from a perfused gland.[4] The Na^+ inward current must be enhancing Ca^{2+} entry by depolarizing the membrane, since the largest Ca^{2+} inward current was recorded under voltage clamp when the membrane was depolarized to $+10$ mV.[15] If this argument is correct blockade of the Na^+ inward current with tetrodotoxin (TTX) should reduce adrenaline secretion induced by ACh. These experiments were carried out in the perfused rat adrenal gland.[23,24] Release of adrenaline was partially blocked by TTX when the ACh concentration was 10 μM or less. This partial inhibition was anticipated. When ACh concentrations are high the depolarization will be large enough to activate a substantial number of voltage-dependent Ca channels without assistance of the Na inward current and also steady depolarization will inactivate some of the Na^+ channels. Thus, the blockade of the Na^+ inward current should not affect the activation of Ca^{2+} channels. Furthermore, Ca^{2+} entering the cell through ACh receptor channels will become significant at high ACh concentrations.

The Ca^{2+} influx through ACh receptor channels was implicated in the following classical experiment. When chromaffin cells were completely depolarized with isotonic K^+-sulfate and then exposed to ACh, additional secretion was observed.[13] Under these conditions ACh does not increase Ca^{2+} entry by further depolarization of the cell. Instead, additional Ca^{2+} ions are likely to enter through ACh receptor channels which are mainly permeable to Na^+ and K^+ ions. ACh causes a slight depolarization even in Na^+-free saline.[4,11,22] In this case inward current is most likely to be carried by Ca^{2+}.

These two major patways for Ca^{2+} contribute to catecholamine secretion. Ca^{2+} entry through other pathways, such as voltage-dependent Na channels, has yet to be evaluated. The next obvious question is the relative contribution of these two major pathways in vivo. Since in the animal ACh is released at the nerve terminal and binds to the receptors clustered at the subsynaptic membrane, which is about 1 μm in diameter,[9] Ca^{2+} entering the cell through ACh receptor channels are limited strictly to this region by the strong Ca^{2+} buffering action of the cytoplasm. On the other hand, voltage-dependent Ca^{2+} channels are probably distributed over the entire surface and Ca^{2+} entry through Ca^{2+} channels could occur anywhere on the cell surface. Thus, upon nerve stimulation Ca^{2+} becomes quickly available for vesicle extrusion in any part of the cell. Some new findings implicate that Ca^{2+} entry through ACh receptor channels is not significantly contributing to catecholamine secretion. Thus, under the voltage clamp, application of ACh did not evoke the membrane capacitance change.[8] The discrepancy between the above-mentioned classical observations by Douglas and Rubin[13] and the recent membrane capacitance measurements can be resolved as described in the "Discussion".

II. CHANGES IN THE ELECTRICAL MEMBRANE CAPACITANCE DURING EXOCYTOSIS

Recently, an exciting avenue was opened by measuring membrane capacitance during exocytosis.[34] It has long been anticipated that there should be a considerable change in the membrane capacitance when the vesicular membrane fuses to the cell surface during exocytosis. This comes from a simple calculation of the capacitance of the vesicular membrane, knowing the size of the vesicle as 0.3 μm in diameter from anatomical data and assuming the specific membrane capacitance of 1 $\mu F/cm^2$. One vesicular membrane should have a capacitance of 2.8 fF (10^{-15} Farad). With the conventional glass microelectrode it is difficult to measure small changes in the capacitance, because the fine microelectrode itself has a large capacitance due to its thin glass wall at the vicinity of the tip. The new technique developed for the single ion channel current recording[17] was required for this measurement. Because of the relatively thick wall and the lower internal resistance of the patch-clamp electrode the electrode capacitance can be well compensated with an appropriate electronic circuit. Then, in the cell-attached whole cell configuration[17] the membrane capacitance of a whole cell can be measured. When a sinusoidal wave of voltage is applied through the electrode the current flowing through capacitative and resistive components can be separated using the difference in the phase relation referring to the original sinusoidal wave. A special amplifier (phase-lock amplifier) was used to separate these two components. With this setup using an 800- to 1600-Hz sinusoidal wave of 5 to 20 mV in amplitude, the background noise with the whole cell configuration was approximately 0.5 to 2 fF (rms, bandwidth 0.1 to 5 Hz) (Figure 1).

Spontaneous capacitative jumps of about 2 fF were observed when the internal solution was strongly buffered at low Ca^{2+} concentration. This was similar to the capacitance change expected when one granule vesicle fused to the surface membrane. When the cytoplasm was weakly buffered for Ca^{2+} (0.1 mM EGTA with 0 Ca^{2+}) and brief depolarization was applied to the membrane, a large increase in the membrane capacitance was observed. However, in the experiment in which the cytoplasm was strongly buffered at a low Ca^{2+} concentration (11 mM EGTA with 1 mM Ca^{2+}, $Ca^{2+} \sim 10$ nM), no capacitative responses were evoked upon depolarization. This observation indicates that even if Ca^{2+} entry occurs, no change in the membrane capacitance takes place when the cytoplasmic Ca^{2+} concentration does not significantly rise. A slow decrease in the membrane capacitance followed a large increase after the depolarizing stimulus. This was considered as the result of a membrane retrieval process.

Having this membrane capacitance measurement technique established an effect of trifluoperazine(TFP) was tested on bovine chromaffin cells.[8] TFP is an inhibitor of calmodulin[28] which is thought to be involved in regulation of various hormone secretion.[25] However, the effect of TFP on catecholamine secretion seems to have multiple sites.[36] Thus, TFP inhibited radioactive Ca^{2+} uptake evoked by carbamylcholine and excess K^+. A precise site of action can be determined by this new technique of membrane capacitance measurement. TFP

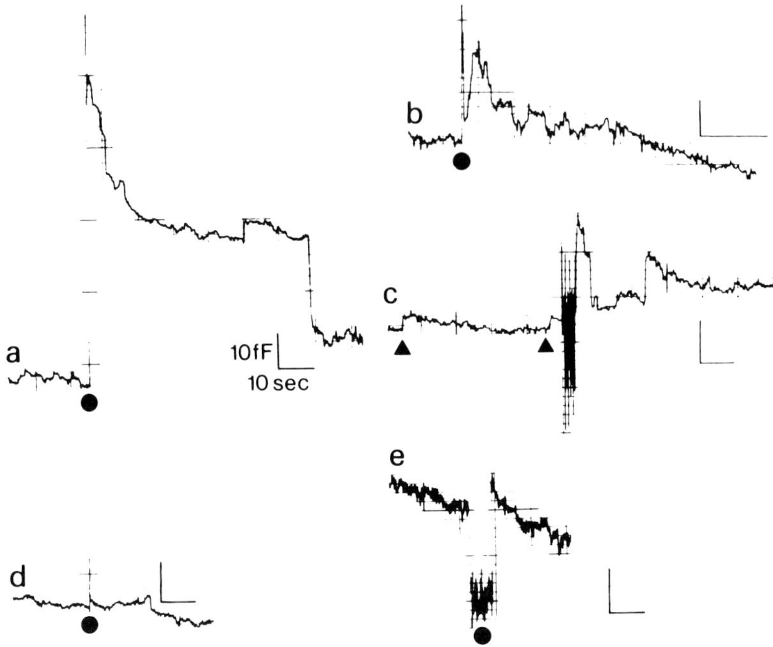

FIGURE 1.   Cell responses to depolarizing potential jumps. The bath solution was 140 mM NaCl, 2.8 mM KCl, 2 mM MgCl$_2$, 1 mM CaCl$_2$ (except in trace e), 10 mM Hepes-NaOH, pH 7.2. The internal (pipette) solution contained 140 mM Na glutamate (traces a, b, and d) or K glutamate (traces c and e), 0.1 mM EGTA (except trace c and d), 2 mM MgCl$_2$, 5 mM ATP, and 10 mM Hepes-NaOH (or KOH) (pH 7.2). The holding potential was -67 mV in traces a, b, and -69 mV in c and e. (All calibration bars = 10 fF and 10 sec.) Traces a and b are typical responses with 0.1 mM EGTA in the pipette solution. In trace a, a 200-msec, 80-mV amplitude depolarization ($\bullet$) elicited an 80-fF C increase, which was followed by a stepwise decrease. The 24-fF rapid decrease at the end of the record appeared on a faster time base to be due to several successive back-jumps. During the 80-mM depolarization, an inward cell current of 60 pA, carried by Ca ions, was observed. Thus, $\sim$3 $\times$ 10^7 Ca ions entered the cell during the depolarizing stimulus. In trace b, the responses to a 70-mV, 200-msec potential stimulation is shown (from another cell). The Ca^{2+}-current amplitude during the stimulus was 40 pA. Trace c shows the response in a cell with 1.5 mM EGTA, 1 mM Ca^{2+} (calculated Ca$^{2+}_i$, 200 nM). $\blacktriangle$, Two spontaneous C jumps of 2 to 3 fF at the beginning of the record. After a 3.5-sec train of depolarizing pulses (10 Hz for 50 msec at 70 mV) a 20-fF C increase occurred, followed by stepwise C changes. Trace d shows the lack of response in a cell with 11 mM EGTA, 1 mM Ca^{2+} (calculated Ca$^{2+}_i$, 10 nM) to an 80-mV, 200-msec voltage stimulation ($\bullet$). Trace e shows the lack of response in a cell bathed in 1 mM Co^{2+} lacking Ca^{2+}. The internal solution contained 0.1 mM EGTA and no Ca (as in traces a and b). A 70-mV, 6-sec-long voltage pulse ($\bullet$) failed to elicit a C signal. The change of the signal during the potential pulse and the stimulus artifacts of c are not necessarily related to actual changes of C for reasons explained in the text. (From Neher, E. and Marty, A., *Proc. Natl. Acad. Sci. U.S.A.*, 79, 6712, 1982. With permission.)

reduced ACh-induced current at the concentration of 0.1 to 1 and 10 μM TFP inhibited voltage-dependent Ca inward current to half of the original amplitude. Finally, 10 μM TFP blocks capacitance changes. The block of capacitance steps by TFP was shown to be independent of the reduction of ACh and Ca inward ionic currents.

This result is in agreement with the observation by Knight and Baker[26] on electrically induced leaky chromaffin cells. However, since the effect of TFP is now shown to be rather nonspecific, TFP might be blocking another yet unknown step or steps in the sequential events leading to hormone secretion. As demonstrated in this paper,[8] the strength of this technique is its ability to separate various steps during stimulus-secretion coupling.

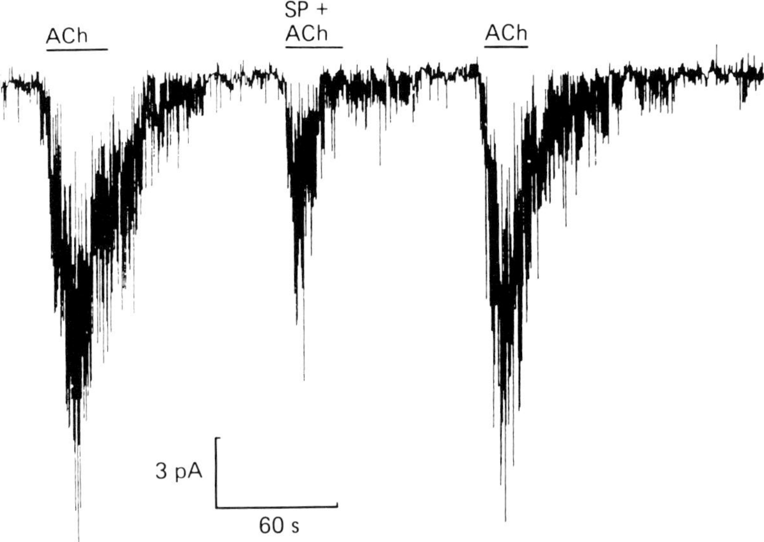

FIGURE 2.   ACh-induced inward current. A single chromaffin cell was voltage clamped to −60 mV, and inward current measured in the presence of 20 μ*M* ACh alone or 20 μ*M* ACh plus 10 μ*M* substance P. Bars denote period of application by bath perfusion. The net current is obviously diminished when substance P is present. Peak current is 13 pA with substance P, compared with 19 pA with ACh alone. Application of agonist solution at a rate of 1 m*ℓ*/min accounts for the slow rise to peak activity. (From Clapham, D. E. and Neher, E., *J. Physiol. (London)*, 347, 255, 1984a. With permission.)

III. DESENSITIZATION OF ACh RECEPTORS AND THE EFFECT OF SUBSTANCE P

While desensitization of ACh receptors in the skeletal muscle has been studied by various investigators,[14,19,35] similar studies in adrenal chromaffin cells are limited.[3,30] Recently, a study using the patch-clamp technique was published.[7] Advantages of using this technique are twofold. The relatively small chromaffin cell can be well voltage-clamped with one electrode and the agonist concentration surrounding the cell can be changed quickly, especially using a special apparatus devised by Krishtal and Pidopulichko.[27] ACh (20 μ*M*) at −60 mV induced an inward current that desensitized in the continued presence of the agonist. The time course of desensitization, in most cases, could be fitted by a single exponential with a time constant of 8 to 10 sec at 22°C (Figure 2). At the single channel current level, bursting and clustering were observed which were similar to ACh receptor channels in the skeletal muscle.[35] Kinetic schemes to explain the behavior of desensitized channels were proposed.

Substance P is probably contained in the splanchnic nerve terminals on chromaffin cells, together with ACh.[29] This raises a possibility that substance P might work as a neuromodulator. Thus, the effect of substance P on ACh-induced release of catecholamines was examined by Clapham and Neher.[7] They tested the effect of substance P on the process of desensitization. It was found that 10 μ*M* substance P accelerates the process of desensitization and correspondingly, the values for channel open time, openings per burst, and bursts per cluster were reduced. The conductance of individual channels was not changed. They concluded that substance P inhibited ACh-induced depolarization of chromaffin cells either by increasing the rate of desensitization or by inducing channel blockade, which apparently enhances desensitization.

In agreement with this finding, Livett et al.[30] found in the dissociated bovine adrenal chromaffin cells that substance P inhibited the ACh-induced catecholamine release. Substance P by itself does not affect the catecholamine release. Inhibition by substance P was non-competitive with ACh. However, an opposite effect of substance P on catecholamine release was also reported by Baksa and Livett.[3] Substance P protected against ACh-induced desensitization. In this experiment ACh receptors were desensitized by incubating cells with high concentrations of nicotine (500 to 3000 μM) for 5 min. The catecholamine release was then tested with 7.5 μM nicotine. In this situation 30 to 50% of catecholamine was released compared with cells which were not incubated with desensitizing concentrations of nicotine. Interestingly, these cells were completely protected from desensitization by 10 μM substance P. Thus, the catecholamine release during the test stimulus was the same as that of non-incubated controls. This result is, at this moment, difficult to reconcile with the previous secretion studies[30] and the electrophysiological findings mentioned earlier.[7]

IV. FURTHER CHARACTERIZATION OF ION CHANNELS

A. Potassium Channels

Marty and Neher[33] identified four types of potassium channels in terms of their unitary currents, i.e., 0.4, 0.8, 1.6, and 5.0 pA at the membrane potential of 0 mV. The 0.8- and 1.6-pA channels were activated by depolarization with a delay and slowly inactivated during prolonged depolarization (delayed rectifier). The 5.0-pA channels are sensitive to internal Ca^{2+} concentration and blocked by 1 mM TEA or 0.1 mM quinine. This type of channel was described earlier by Marty[31] in the chromaffin cell (Ca^{2+}-dependent potassium channels, Figure 3). Interestingly, this channel can be blocked by internal Na^+ ions. Marty[32] found that internal 20-mM Na^+ induced short interruptions of the outward K^+ current in a voltage-dependent manner. In addition, internal Na^+ reduced the opening probability of the channels. The blocking effects of internal Na^+ were relieved by increasing the external K^+ concentration. These findings suggest that the channels possess, in addition to the Ca^{2+}-binding sites presumably involved in channel activation,[2] internally located Na^+-binding sites and externally located K^+-binding sites which are probably both closely associated with the permeation pathway. The binding of K^+ to the latter sites appears to displace the Na^+ internally bound, perhaps as a result of electrostatic repulsion.

These are interesting observations which may give us hints of the molecular mechanism of ion permeation. How these channels contribute to the function of the hormone-secreting chromaffin cell has yet to be elucidated.

B. ACh Receptor Channels

In order to understand the activation characteristics of ACh receptors, the relationship between the ACh concentration and the ACh-induced current is important. Previously, the relationship between the amplitude of ACh-induced depolarization and the ACh concentration was studied by applying known concentrations of ACh with the "puffer" technique.[22] The depolarization became detectable at 0.1 μM ACh and saturated at around 100 μM. The half-maximal concentration was at about 10 μM. This range of effective ACh concentrations is similar to that necessary to stimulate adrenaline secretion from the perfused rat adrenal medulla.[18,22] There are, however, three major problems in this dose-response curve to obtain information on channel activation characteristics. First, the depolarization evoked by high concentrations of ACh almost reached the apparent reversal potential, and thus ceiling. Second, various types of ion channels are presumably contributing to ACh-induced depolarization. Besides ACh receptor channels, voltage-dependent channels are likely to be activated by the depolarization. Further Ca^{2+} ions that enter the cell through ACh receptor channels and through voltage-dependent calcium channels may activate Ca^{2+}-dependent potassium channels[31] and nonselective cation channels.[16] Third, desensitization of ACh receptor channels may prohibit their complete activation.

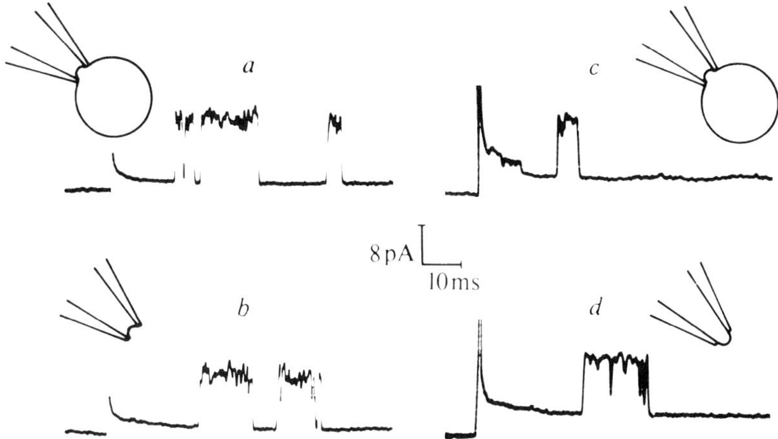

FIGURE 3.   Large K unitary currents in isolated patches. (a, b) A pipette containing Ca-free Ringer (1 m*M* Co replacing Ca) was first used to record from a patch of a cell membrane (a). The bath contained a high-K solution (in m*M*: 143 KCl, 2 MgCl$_2$, 1 $^{-11}$Ca EGTA buffer, 10 Hepes-KOH, pH 7.3) which presumably brought the cell membrane potential close to 0 mV. The pipette was held at the bath potential, so that the holding potential for the patch was also 0 mV. (a) 75-mV potential jumps were applied to the pipette interior to make the patch membrane potential positive; (b) response to the same voltage pulse after moving the pipette away from the cell (inside-out patch). Unitary outward currents are 12 pA in a, 10.5 pA in b. There is a pronounced noise increase during the opening of the channels. (c, d) A pipette containing a high-K solution (same as the bath solution in a, b) was first applied to a cell bathed in normal Ringer (c). The pipette interior was held at the bath potential so that the patch-holding potential was equal to the cell-resting potential (presumably, ~ -60 mV). 100-mV Positive pulses activate large (11 pA) K unitary currents; (d) similar currents are obtained after destruction of the initial patch and withdrawal of the pipette when applying 104-mV pulses from a holding potential of -64 mV (outside-out patch). 1-kHz filtering. (From Marty, A., *Nature (London)*, 291, 497, 1981. With permission.)

In an attempt to determine a more accurate dose-response relation, Kidokoro et al.[21] voltage clamped the cultured rat chromaffin cell and applied various concentrations of ACh with the "puffer" method. The voltage clamping eliminates the first and a part of the second problem. The remaining second problem was dealt with by using Cs$^+$ in the internal solution to eliminate current flowing through K channels. The most difficult one was the third problem which remained, making the dose-response curve at high concentrations unreliable. Nevertheless, the ACh-current relation thus obtained is useful. The dissociation constant was estimated as 75 to 170 μM and the Hill coefficient was 1.2 to 1.5. The apparent Kd was much higher than that estimated previously and is probably still an underestimate. It appears that binding of more than one ACh molecule is required for activation of ACh receptors in chromaffin cells, as was found in the skeletal muscle.[1,5,10]

V. DISCUSSION

Since the development of the patch-clamp technique the chromaffin cell became a subject of detailed biophysical investigations. The molecular mechanism of ion permeation and membrane fusion process can be suitably studied in this system. A cleanly separated and relatively homogeneous population of these cells in primary cultures is now an advantage for various electrophysiological investigations as well as biochemical works. Undoubtedly, this trend will continue and will provide us with invaluable information on the membrane properties of chromaffin cells. However, nagging questions as to how these cells behave *in situ* still remain. For example, ACh receptor channels studied in culture are most likely to be newly synthesized ones after isolation of cells. It is well known in the skeletal muscle that these new receptors appearing after denervation are different from those at the subsynaptic area. Whether that is also true in the chromaffin cell has to be determined. The role of electrical coupling cannot be studied in culture. It is quite important to resume electrophysiological studies of cells *in situ* in parallel with more detailed biophysical analyses in culture.

New pieces of information are accumulating regarding the events occurring in the vicinity of the surface membrane. It seems worth trying to explain some previously perplexing observations. ACh causes a slight depolarization, even in Na^+-free saline.[4,11,22] In this case inward current is most likely to be carried by Ca^{2+}. If this is the case, ACh should stimulate adrenaline release even in Na-free saline, but in the majority of cases no secretion was observed in cultured bovine chromaffin cells after long exposure to Na-free solutions.[6,37] This phenomenon was difficult to explain until recently, when a new piece of information was published by Clapham and Neher.[8] Upon application of 50 μM ACh membrane conductance increased transiently under voltage-clamped conditions. However, no significant change in membrane capacitance was recorded. Clapham and Neher[8] concluded that apparently the Ca^{2+} entry through ACh receptor channels is not sufficient to elicit exocytosis. It is clear that some Ca^{2+} enters the cell through ACh receptor channels. The secretion mechanism senses the Ca^{2+} concentration at the immediate vicinity of the surface membrane. If the concentration does not reach the threshold the hormone secretion does not take place. Therefore, depending on the condition of cells, such as completely depolarized ones with isotonic K-sulfate[12] where presumably internal Ca^{2+} concentrations are higher than at the normal state, small Ca^{2+} influx may trigger the release of catecholamine.

Since in the animal ACh receptors are clustered in the subsynaptic membrane and since Ca^{2+} ions are strongly buffered in the cytoplasm, Ca^{2+} ions entering the cell through these channels are probably limited strictly to this small region.

In view of these findings, Ca^{2+} entry through ACh receptor channels may not significantly contribute to catecholamine secretion.

REFERENCES

1. **Adams, P. R.,** An analysis of the dose-response curve at voltage-clamped frog endplates, *Pflugers Arch.,* 360, 145, 1975.
2. **Barret, J. N., Magleby, K. L., and Pallota, B. J.,** Properties of single calcium-activated potassium channels in cultured rat muscle, *J. Physiol. (London),* 331, 221, 1982.
3. **Boksa, P. and Livett, B. G.,** Substance P protects against nicotinic desensitization of cultured adrenal chromaffin cells, *Soc. Neurosci. Abstr.,* 8, 986, 1982.
4. **Brandt, B. L., Hagiwara, S., Kidokoro, Y., and Miyazaki, S.,** Action potentials in the rat chromaffin cells and effects of acetylcholine, *J. Physiol. (London),* 263, 417, 1976.
5. **Catterall, W. A.,** Sodium transport by the acetylcholine receptor of cultured muscle cells, *J. Biol. Chem.,* 250, 1776, 1975.
6. **Cena, V., Nicolas, G. P., Sanchez-Garcia, P., and Garcia, A. G.,** Limitation of the pharmacological approach to dissect ionic channels involved in catecholamine release from chromaffin cells, in Molecular Neurobiology of Pheripheral Catecholaminergic System, Abstr., Ibiza, Spain, 1982.
7. **Clapham, D. E. and Neher, E.,** Substance P reduces acetylcholine-induced currents in isolated bovine chromaffin cells, *J. Physiol. (London),* 347, 255, 1984a.
8. **Clapham, D. E. and Neher, E.,** Trifluoperazine reduces inward ionic currents and secretion by separate mechanisms in bovine chromaffin cells, *J. Physiol. (London),* 353, 541, 1984b.
9. **Coupland, R. E.,** Electron microscopic observations on the structure of the rat adrenal medulla. I. The ultrastructure and organization of chromaffin cells in the normal adrenal medulla, *J. Anat.,* 99, 231, 1965.

10. **Dionne, V. E., Steinbach, J. H., and Stevens, C. F.,** An analysis of the dose-response relationship at voltage-clamped frog neuromuscular junctions, *J. Physiol. (London),* 281, 421, 1978.
11. **Douglas, W. W., Kanno, T., and Sampson, S. R.,** Effects of acetylcholine and other medullary secret-agogues and antagonists on the membrane potential of adrenal chromaffin cells: an analysis employing techniques of tissue culture, *J. Physiol. (London),* 188, 107, 1967.
12. **Douglas, W. W. and Rubin, R. P.,** The role of calcium in the secretory response of the adrenal medulla to acetylcholine, *J. Physiol. (London),* 159, 40, 1961.
13. **Douglas, W. W. and Rubin, R. P.,** The mechanism of catecholamine release from the adrenal medulla and the role of calcium in stimulus-secretion coupling, *J. Physiol. (London),* 167, 288, 1963.
14. **Feltz, A. and Trautmann, A.,** Desensitization at the frog neuromuscular junction: a biphasic process, *J. Physiol. (London),* 322, 257, 1982.
15. **Fenwick, E. M., Marty, A., and Neher, E.,** A patch-clamp study of bovine chromaffin cells and of their sensitivity to acetylcholine, *J. Physiol. (London),* 331, 577, 1982a.
16. **Fenwick, E. M., Marty, A., and Neher, E.,** Sodium and calcium channels in bovine chromaffin cells, *J. Physiol. (London),* 331, 599, 1982b.
17. **Hammill, O. P., Marty, A., Neher, E., Sakmann, B., and Sigworth, F. J.,** Improved patch-clamp techniques for high-resolution current recording from cells and cell-free membrane patches, *Pflugers Arch.,* 391, 85, 1981.
18. **Ishikawa, K., Harada, E., and Kanno, T.,** A quantitative analysis of the influence of external calcium and magnesium concentrations on acetylcholine-induced adrenaline release in rat adrenal gland perfused in an antidromic or orthodromic direction, *Jpn. J. Physiol.,* 27, 251, 1977.
19. **Katz, B. and Thesleff, S.,** A study of the desensitization produced by acetylcholine at the motor end-plate, *J. Physiol. (London),* 138, 60, 1957.
20. **Kidokoro, Y.,** Electrophysiology of adrenal chromaffin cells, in *The Electrophysiology of the Secretory Cell,* Poisner, A. and Trifaró, J. M., Eds., Elsevier, Amsterdam, 1985, chap. 8.
21. **Kidokoro, Y., Hirano, T., and Ohmori, H.,** Acetylcholine dose-response relationship in the rat adrenal chromaffin cell, in Int. Symp. on the Molecular Biology of Catecholamine Storing Tissues, Colmar, France, 1984, 74.
22. **Kidokoro, Y., Miyazaki, S., and Ozawa, S.,** Acetylcholine-induced membrane depolarization and po-tential fluctuations in the rat adrenal chromaffin cells, *J. Physiol. (London),* 324, 203, 1982.
23. **Kidokoro, Y. and Ritchie, A. K.,** Chromaffin cell action potentials and their possible role in adrenaline secretion from rat adrenal medulla, *J. Physiol. (London),* 307, 199, 1980.
24. **Kidokoro, Y., Ritchie, A. K., and Hagiwara, S.,** Effect of tetrodotoxin on adrenal secretion in perfused rat adrenal medulla, *Nature (London),* 278, 63, 1979.
25. **Klee, C. B., Crouch, T. H., and Richman, P. G.,** Calmodulin, *Ann. Rev. Biochem.,* 49, 489, 1980.
26. **Knight, D. E. and Baker, P. F.,** Calcium-dependence of catecholamine release from bovine adrenal medullary cells after exposure to intense electric fields, *J. Membr. Biol.,* 68, 107, 1982.
27. **Krishtal, O. A. and Pidoplichko, V. I.,** A receptor for protons in the cell membrane, *Neuroscience,* 5, 2325, 1980.
28. **Levin, R. M. and Weiss, B.,** Binding of trifluoperazine to the calcium-dependent activation of cyclic nucleotide phosphodiesterase, *Mol. Pharmacol.,* 13, 690, 1977.
29. **Linnoila, R. I., Diaugustine, R. P., Hervonen, A., and Miller, R. J.,** Distribution of (Met[5])- and (Leu[5])-enkephalin-vasoactive intestinal-polypeptide, and substance P-like immuno-reactivities in human adrenal glands, *Neuroscience,* 5, 2247, 1980.
30. **Livett, B. G., Kozousek, V., Mizobe, F., and Dean, D. M.,** Substance P inhibits nicotinic activation of chromaffin cells, *Nature (London),* 278, 256, 1979.
31. **Marty, A.,** Ca-dependent K channels with large unitary conductance in chromaffin cell membranes, *Nature (London),* 291, 497, 1981.
32. **Marty, A.,** Blocking of large unitary calcium-dependent potassium currents by internal sodium ions, *Pflugers Arch.,* 396, 176, 1983.
33. **Marty, A. and Neher, E.,** Potassium channels in bovine adrenal chromaffin cells, in Int. Symp. on the Molecular Biology of Catecholamine Storing Tissues, Colmar, France, 1984, 76.
34. **Neher, E. and Marty, A.,** Discrete changes of cell membrane capacitance observed under conditions of enhanced secretion in bovine adrenal chromaffin cells, *Proc. Natl. Acad. Sci. U.S.A.,* 79, 6712, 1982.
35. **Sakmann, B., Patlak, J., and Neher, E.,** Single acetylcholine-activated channels show burst kinetics in the presence of desensitizating concentration of agonist, *Nature (London),* 286, 71, 1980.
36. **Wada, A., Yanagihara, N., Izumi, F., Sakurai, S., and Kobayashi, H.,** Trifluoperazine inhibits $^{45}Ca^{2+}$ uptake and catecholamine secretion and synthesis in adrenal medullary cells, *J. Neurochem.,* 40, 481, 1983.
37. **Wada, A., Yashima, N., Izumi, F., Kobayashi, H., and Yanagihara, N.,** Involvement of Na influx in acetylcholine receptor mediated secretion of catecholamines from cultured bovine adrenal medulla cells, *Neurosci. Lett.,* 47, 75, 1984.

Chapter 14

MODULATION BY CALCIUM OF THE KINETICS OF THE CHROMAFFIN CELL SECRETORY RESPONSE

A. G. García, F. Sala, V. Ceña, C. Montiel, and M. G. Ladona

TABLE OF CONTENTS

I. INTRODUCTION

In this chapter, we seek to provide recent information relating the modulation by Ca of the adrenomedullary catecholamine release response. Such a response is physiologically triggered when an action potential invades the splanchnic nerve terminals innervating chromaffin cells; the released acetylcholine[1,2] causes the exocytotic secretion of catecholamines by stimulating cholinoceptors located at the chromaffin cell membrane.

In 1961, Douglas and Rubin[3] concluded that "the role of acetylcholine as a transmitter at the adrenal medulla is to cause some brief change in medullary cells which allows extracellular Ca (Ca_o) to penetrate them and trigger the catecholamine ejection process". Since then, information has been accumulated concerning the elucidation of the possible pathways used by Ca to gain access to the secretory machinery, the kinetics of voltage-dependent Ca channels, and the estimation of intracellular-free Ca levels (Ca_i) at resting and stimulating conditions. Some of this information has been sought through the direct measurement of radiolabeled Ca ion transients, monitoring of its intracellular levels by using leaky cells and Ca-sensitive fluorophores, or sophisticated electrophysiological techniques (i.e., patch-clamp recordings of single Ca channel currents). A major breakthrough providing additional valuable information comes from the discovery of new pharmacological tools (novel dihydropyridines Ca channel "activators") that are allowing the fine kinetic analysis of currents carried out through the Ca channel; they are also stimulating initial attempts to isolate, purify, and characterize the transmembranous proteic complex associated with the channel. The discovery of these interesting new drugs has also led to the suggestion that an endogenous substance (a peptide of low molecular weight?) might be manufactured by the chromaffin cell to modulate the activity of Ca channels.

Using these new methodologies and pharmacological tools, data have been obtained that allow to draw a clearer picture concerning the sequence of events taking place during the secretory cycle. Briefly, acetylcholine released at the splanchnic nerve-chromaffin cell junction binds to its receptor and opens the associated ionophore, allowing Na entry and depolarization of the excited chromaffin cell membrane. Depolarization will then activate adjacent voltage-sensitive Ca channels, producing Ca entry and secretion. Catecholamine release is then triggered by a critical rise of Ca_i. Because Ca_o seems to gain the cell interior primarily through potential-sensitive Ca channels, their rates of activation and inactivation are likely to limit the rates of Ca access to the cell and, obviously, this may constitute an efficient control system to modulate the physiologiccal catecholamine release process. An attempt to correlate catecholamine release and Ca transients, and a second one directed to define the relative importance of a membrane-bound vs. a cytosolic mechanism responsible for the transient nature of the catecholamine release response upon sustained depolarization of the chromaffin cell, constitute the two main objectives of this article.

II. PATHWAYS FOR CALCIUM ENTRY DURING THE SECRETORY CYCLE

Several data from our laboratory suggest that acetylcholine, nicotine, or high K all activate Ca entry into the chromaffin cell (and the subsequent catecholamine release) chiefly via a common voltage-dependent Ca channel.[4-6] However, it has also been suggested that Ca may enter the cell through a second pathway, the nicotinic receptor-associated ionophore.[7-9] Quantitation of the degree of participation of an ionic channel in allowing Ca entry during secretion has been based in two types of studies: (1) determination of the effects of drugs known to act on specific ionic channels on catecholamine release, ionic movements, or electrical properties of chromaffin cells; and (2) observations on how certain specific manipulations of bathing media affected such parameters. Several experimental approaches have been used to determine the relative contribution of each channel to the entry of Ca inside chromaffin cells.

Acetylcholine increases catecholamine secretion from perfused cat[10] and rat adrenal glands[7] in the presence of high K, a situation in which the chromaffin cell is already depolarized. These experiments suggest that high K depolarization might activate voltage-dependent Ca channels, while cholinergic stimulation, which appears to be partly independent of depolarization-evoked Ca entry, might facilitate Ca entry through a different receptor-operated Ca channel. However, these experiments were done using acetylcholine, which is known to cause the release of catecholamines from cat adrenal glands by activation of both nicotinic and muscarinic receptors.[11,12] When muscarine is used in the presence of high K, its secretory response is preserved;[12] on the contrary, the nicotine response in high K is abolished in our hands.[5] So, the secretory response to nicotine in the perfused adrenal gland can be distinguished from the acetylcholine response by its inability to potentiate the late secretory response to high K. In the light of these results, the strength of earlier experiments of Douglas and Rubin[10] and Ishikawa and Kanno[7] referred to earlier, suggesting that Ca enters the chromaffin cell through the nicotinic receptor ionophore to evoke catecholamine release, is weakened because acetylcholine (which has nicotinic as well as muscarinic effects) was used as secretagogue.

If both nicotine or high K cause the release of catecholamines by increasing a common Ca permeability pathway (i.e., the voltage-sensitive Ca channel), the inorganic Ca entry blocker Mn or the organic Ca antagonist nitrendipine should equally inhibit secretion in response to both secretagogues. Mn inhibited catecholamine release evoked by acetylcholine, nicotine, or high K from perfused cat adrenal glands in a similar manner.[5] In addition, nitrendipine blocked equally ^{3}H-noradrenaline release from cultured bovine adrenal chromaffin cells evoked by 5 μM nicotine or 59 mM K.[4] Conversely, the Ca channel activator BAY-K-8644 potentiated similarly the secretion evoked by nicotine or high K in perfused cat adrenal glands (Figure 1), with a similar time course of the secretory profile.

All these experiments do not invalidate the contention that some Ca might enter the chromaffin cell through the acetylcholine receptor ionophore during the physiological catecholamine release process; however, they do support our belief that during nicotinic or high K-depolarization Ca enters the cell primarily through potential-dependent specific Ca channels. What is known about the pharmacology and the kinetics of these channels will be described next.

III. THE CHROMAFFIN CELL CALCIUM CHANNEL

Several approaches measuring catecholamine release, ^{45}Ca fluxes, binding of radioligands, or electrical recordings have been used to demonstrate that the chromaffin cell contains Ca channels that seem to play a major role in stimulus-secretion coupling. This is suggested on the grounds that catecholamine release evoked by several secretagogues is inhibited by agents that are presumed to act selectively on such channels. This is true for some divalent cations (Mg, Mn, Co) or the so-called Ca channel antagonists. A growing number of these drugs known to block voltage-sensitive Ca channels in different tissues are available.[13] They constitute a heterogeneous group with disparate chemical structures, pharmacological effects, and clinical profiles. Within their overall classification, they have been subclassified in subgroups with separate molecular sites of action on the transmembranous macromolecular Ca channel complex. Although the classifications vary from one to another author, they are generally grouped in three classes:[14] class I, dihydropyridines (nifedipine, PY 108-068, nimodipine, nitrendipine, niludipine, nisoldipine); class II (verapamil, D-600, diltiazem, diclofurime); and class III, diphenylalkylamines (pimozide, fendiline, lidoflazine, perhexiline, cinnarizine, prenylamine, flunarizine).

One of the main objections to the use of these drugs to characterize Ca channels is their limited selectivity and their relative efficacy to block Ca channels from neurones and other

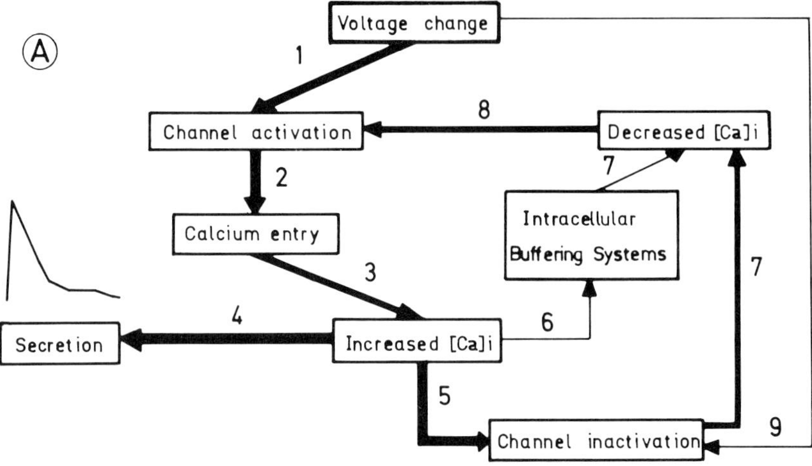

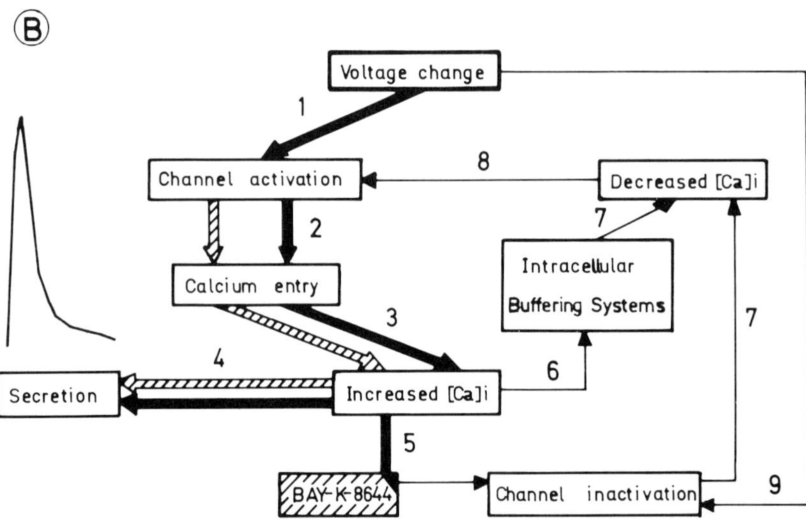

FIGURE 1.   Scheme showing the possible relationship between the extent of catecholamine release responses and the kinetics of activation and inactivation of chromaffin cell potential-dependent Ca channels during perfusion of cat adrenal glands with Krebs solutions containing high K concentrations. (A) *In the absence of BAY-K-8644*. High K causes cell depolarization and the activation (1) of voltage-sensitive Ca channels; Ca enters the cell (2) following a large electrochemical gradient, producing an increase of Ca_i (3) that triggers the exocytotic release of catecholamines (4). Both depolarization (9) and enhanced Ca_i (3) cause the inactivation of Ca channels (5); then Ca entry is decreased. Because intracellular Ca is sequestered by buffer systems (6), Ca_i is lowered (7). This decrease of Ca_i is greatly accelerated by Ca_o omission or Co addition; in the presence of sustained depolarization, low Ca_i will reactivate the channel (8) in such a manner that Ca reintroduction or Co omission will allow Ca entry and renewal of the secretory response. (B) *In the presence of BAY-K-8644*. Assuming that the dihydropyridine Ca channel activator BAY-K-8644 (and a dihydropyridine-like endogenous modulator in physiological conditions) binds to an inactivation site located intracellularly on the transmembraneous Ca channel, and that this site closes the channel by some reaction that is triggered by the binding of Ca, the inactivation process referred to in A will be delayed (5); then, initially the channels will undergo long-lasting openings, allowing more Ca entry and a potentiated catecholamine secretory response (dashed arrows). In a second phase, Ca_i will be markedly increased (3), and this excess Ca will activate the inactivating site and counteract the effects of BAY-K-8644 at its channel binding site; eventually, inactivation of the channel will then take place and also the secretory response (4) will decline.

secretory cells. So, verapamil, the first Ca entry blocker reported to depress potently cardiac tissues,[15] inhibits catecholamine release in response to acetylcholine or high K only at concentrations several orders of magnitude greater than those required to affect heart or smooth muscle tissues.[16-18] In addition, verapamil also blocks Na channels in synaptosomes,[19] has muscarinic cholinoceptor[20] and α_2-adrenoceptor affinity[21] and also inhibits Ca conductance of acetylcholine endplate channels.[22]

The more recent organic Ca antagonists of the dihydropyridine type seem to be more potent and specific at neuronal cell types. They inhibit at the nanomolar range the release of ^{3}H-noradrenaline from PC12 cells;[23] in our hands, nitrendipine also inhibited the release of ^{3}H-noradrenaline evoked by nicotine or high K from bovine adrenomedullary chromaffin cells with IC50s of 400 and 20 nM, respectively.[4] In contrast, 0.5 mM verapamil was needed to block high K-evoked catecholamine release from rat adrenal glands by 75%[18] and 10 μM verapamil or D600 to inhibit the acetylcholine-evoked release from isolated bovine adrenal chromaffin cells.[17] Since verapamil and dihydropyridines are equally effective in blocking heart Ca channels, their different potencies to affect adrenomedullary catecholamine release suggest that the chromaffin cell might contain two different types of Ca channels with different pharmacological properties.

A second evidence favoring the presence in the chromaffin cell of specific Ca channels comes from the fact that both nicotinic stimulation or high K greatly increase the rate of ^{45}Ca uptake by the adrenal medulla[24] or isolated chromaffin cells[8,9] in a manner parallel to the secretion of catecholamines. This ^{45}Ca-uptake increase was abolished when inorganic Ca-channel blockers were used.

Measurements of ^{3}H-nitrendipine-specific binding to membrane fragments from bovine adrenal medulla demonstrated a saturable binding with an apparent K_D of 1.18 $\pm$ 0.32 nM and a B_{max} of 325 $\pm$ 136 fmol/mg protein.[25] These values are compatible with the presence in the chromaffin cell membrane of a high affinity binding site probably related to the voltage-sensitive Ca channel. Although the binding of the drug is not substantially affected by Ca ions, it is worth noting that the Ca chelating agent EGTA drastically inhibited ^{3}H-nitrendipine binding and Ca ions did antagonize this inhibition, suggesting that some small membrane-bound Ca pool, structurally related to the Ca channel complex, might be required for the binding of the organic Ca antagonist to its receptor. On the contrary, drugs acting on the acetylcholine site (nicotine) or the ionophore (imipramine, cocaine) of the nicotinic cholinoceptor did not affect the binding of ^{3}H-nitrendipine, suggesting that indeed this ligand is bound to an extrareceptor site, most likely the voltage-sensitive Ca channel of the chromaffin cell membrane. If nitrendipine binds only to Ca channels and in a 1:1 fashion, these data suggest approximately a density of one to ten channels per square micrometer (calculated from our data[25] by Baker[26]).

Electrical recordings from chromaffin cells have also added evidence on the characteristics of their Ca channels. Acetylcholine or high K were shown to depolarize chromaffin cells by recording the mean value of the transmembrane potential in chromaffin cell cultures of gerbils[27,28] and rats[29-31] and in intact perfused rat adrenal glands.[7] The membrane potential fell progressively as the K concentration was raised and there was an almost linear relationship between the logarithm of the K concentration, the membrane potential, and the rate of adrenaline release, supporting the view that depolarization induced by excess K causes the influx of Ca ions, which initiates the exocytotic event.[7]

A D600 and cobalt-sensitive Ca current has been described in adrenal medullary cells and the underlying unit conductance characterized.[32] The current is maximal in the region of zero potential and has a reversal potential close to $+60$ mV, not far enough positive for a pure Ca current. This suggests that an intracellular cation, perhaps K, may be able to move outwards through the channel. Both currents and unit conductances have proved to be larger and easier to study when Ca is replaced as the major cation by Ba. Following a depolarizing step, Ca channels tend to open in bursts separated by long, silent periods. Of particular interest is the finding that the rate of channel openings only declines very slowly (over many seconds) after application of a steady depolarization. This is somewhat surprising because both the rise in free Ca and associated secretion evoked by K depolarization are transient.[33,60] It may be that the Ca channels are inactivated by a rise in internal Ca[34,35] and that this is minimized under the conditions of patch clamp that have been used so far; but there would appear to be a discrepancy between patch clamp data and intact cells that awaits resolution.[26] Estimates of the number of Ca channels with patch-clamp recording techniques[32] suggest 5 to 15/μm^2 a figure somewhat higher, but in the same general range as that obtained from ^{3}H-nitrendipine binding studies.[25]

IV. CALCIUM CHANNEL ACTIVATORS

The so-called Ca channel agonists or activators represent another particularly exciting recent pharmacological development. Small modifications to the nifedipine molecule produced a derivative, BAY-K-8644 (methyl-1,4-dihydro-2,6-dimethyl-3-nitro-4-[2-trifluoro-methylphenyl]-pyridine-5-carboxylate), that in contrast to the Ca channel blocking agents stimulated cardiac and vascular smooth muscle contractility.[36] In contrast to the ionophores A-23187[37] or ionomycin[4,38] this drug did not enhance by itself the liberation of catecholamines from the perfused cat adrenal gland, but potentiated the release evoked by Ca reintroduction, ouabain, moderate high K concentrations, or nicotine (Figure 1); also, BAY-K-8644 increased K-evoked ^{45}Ca uptake by isolated bovine adrenal chromaffin cells and displaced the specific binding of ^{3}H-nitrendipine to bovine adrenomedullary membrane fragments.[25] The potentiating effects of BAY-K-8644 on catecholamine release of ^{45}Ca uptake are antagonized by nitrendipine or nifedipine, suggesting that both dihydropyridine agonists and antagonists act on the same site at the Ca channel.[25,39] These initial data caused a great expectation, because BAY-K-8644 seemed to arise as the first compound available to manipulate pharmacologically the rates of activation or inactivation of Ca channels.[40,41] In fact, the drug is proving to be an excellent and useful tool to study the properties of such channels, as well as their functional dynamics during the secretory cycle for catecholamines and other neurotransmitters and hormones. We will now describe in more detail experiments recently performed in our laboratory in this direction.

Two recent experimental designs performed in our laboratory strongly support our belief that BAY-K-8644 is specifically acting on chromaffin cell Ca channels to prolong their opening times, thereby delaying their inactivation rates. First, the ionophore A23187 (a drug that increases the rate of spontaneous catecholamine release by directly introducing Ca into the chromaffin cell, by-passing the Ca channel[37]) gave secretory responses proportional to Ca$_o$ concentrations. Such responses were not potentiated at all by BAY-K-8644.[41a]

The second experimental design deals with other divalent cations (Sr and Ba) that can substitute for Ca in releasing catecholamines from cat adrenal glands.[42] The relative conductances and permeability ratios of Ca channels for Ca, Sr, and Ba are known to be quite different in several neuronal cell types;[43-46] in fact, Ba currents are larger than Sr and Ca currents at the bovine chromaffin cell.[47]

Selecting extracellular divalent cation concentrations that give a similar quantitative secretory response when applying brief high-K depolarizing pulses, we have recently demonstrated at the perfused cat adrenal gland that BAY-K-8644 markedly potentiated the release peaks when using Ca; in the presence of Sr, the potentiation observed was modest, and with Ba it was absent. These data support previous experiments from our laboratory monitoring ^{45}Ca^{2+} fluxes and ^{3}H-nitrendipine binding[25] and strongly suggest that dihydropyridines modulate catecholamine release by acting selectively on membrane Ca channels. The question arises now on how the Ca channel activity itself is modulated by these drugs and whether from this pharmacological approach a model for the functioning of Ca channels during the secretory cycle can be established.

In our hands, BAY-K-8644 is able to enhance secretory responses more effectively as the smaller size of the stimulus used is (i.e., moderate high K_o or Ca_o concentrations): there is no potentiation when using strong stimuli (59 mM K in the presence of 2.5 mM Ca). This agrees with data measuring Ca channel current with a whole-cell suction pipette recording from a guinea pig myocyte; BAY-K-8644-induced enhancement of peak current was largest at weak depolarizations (where the current was small in the absence of drug) and became progressively smaller with increasing depolarizations.[40]

From these results it is likely to think that BAY-K-8644 is acting in a stimulus- and Ca_i-dependent manner. This is compatible with the hypothesis implicating Ca_i as responsible for Ca-channel inactivation.[34,35] If Ca_i promotes inactivation of the channel through the binding to an intracellular inactivation site on the transmembranous channel, BAY-K-8644 may bind to a second site (a dihydropyridine receptor?) that mediates the kinetics of the inactivation process. If such receptor exists, it is likely that an endogenous dihydropyridine-like agonist for such a receptor might be manufactured by the cell to control, in a coordinated manner with Ca_i, the activation and inactivation of Ca channels. This interpretation is in line with the recent findings of Hess et al.,[40] Koubun and Reuter,[41] and Brown et al.[48] that BAY-K-8644 promotes long-lasting openings and very brief closings in cardiac cells.

If there is an inactivating (closing) mechanism sensitive to Ca on Ca channels, their mean open times will be longer (as it occurs in the presence of BAY-K-8644) when the permeable cations are different (Sr or Ba)[43-47] from Ca, and BAY-K-8644 should enhance these currents in different degrees according to the permeabilities and conductances of Ca channels for the divalent cation used. In fact, in our hands BAY-K-8644 potentiated markedly the high K evoked secretion in the presence of Ca, little the release evoked in the presence of Sr, and nothing at all the release evoked in the presence of Ba as discussed previously.

A scheme showing how BAY-K-8644 may contribute to the enhancement of the secretory response is represented in Figure 1.

V. KINETICS OF THE ACTIVATION AND INACTIVATION OF CATECHOLAMINE RELEASE

Upon prolonged stimulation of the cat adrenal gland with high K or acetylcholine, the amount of catecholamine release into the perfusate declines exponentially during the first few minutes and falls to about 10% of the initial rate by the tenth minute.[3,10] The cause for this reduction in catecholamine output is not well understood; the rate-limiting step might be located at the plasma membrane (desensitization of the acetylcholine receptor, inactivation of the Ca channel) or at an intracellular site (refractoriness of the exocytotic mechanism, Ca buffering, or Ca extruding systems). We will analyze separately the inactivation of the nicotinic and high K responses.

VI. THE SECRETORY RESPONSE TO PROLONGED STIMULATION WITH NICOTINE

The analysis of the inactivation of the secretory response during cholinergic stimulation of the adrenal gland has the potential difficulty in distinguishing between the contributions of acetylcholine receptor desensitization and Ca channel inactivation. It is assumed, however, that the decline of responsivity during cholinergic stimulation is chiefly related to desensitization of the nicotinic receptor of the chromaffin cell.[49] The nicotinic response has a profile similar to the high K response and both may involve similar mechanisms. The two responses have similar time courses at the higher concentrations used and for both secretagogues the onset of secretion tends to be delayed at the lower concentrations[49] (Figure 2). In addition, BAY-K-8644 caused a great potentiation of the secretory response to low concentrations of nicotine or K, leaving unaffected the secretory profile obtained with higher nicotine or K concentrations. The reactivation by Ca deletion and its subsequent reintroduction of the secretory response to nicotine or high K were also similar in both processes, suggesting that a common mechanism might be involved in both.

However, some data suggest that secretory responses to nicotine and high K are inactivated by different mechanisms; one such difference is the fact that upon repetitive nicotine or high K stimulation of cat adrenal glands, the secretory responses to nicotine decline very substantially, and their recovery is very slow; with high K the extent of such responses is quantitatively similar upon repetitive stimulation. A second difference is the fact that the adrenal response to high K and normal Ca is moderately reduced after 10 min pretreatment with high K in Ca-free medium, but 10 min pretreatment with nicotine in Ca-free medium does not diminish the adrenal response to nicotine and normal Ca.[49] These two distinctions are consistent with the possibility that nicotine-evoked secretion declines as a consequence of receptor desensitization; that Ca plays a major role in this process facilitating desensitization, as in other cholinergic receptors at the skeletal muscle,[50,51] is proven in the experiment shown on Figure 2A and B, where Ca removal and its subsequent reintroduction restores the desensitized catecholamine secretory responses evoked by sustained stimulation with nicotine. These experiments suggest that reduction of responsiveness due to an increase in intracellular Ca is responsible for the inactivation of the secretory response to nicotinic stimulation.

However, the fact that BAY-K-8644 potentiated equally the secretory response to nicotine or high K suggests that the Ca channel is involved in the secretory response to both processes; if so, it seems reasonable to conclude that such channel might also contribute to the inactivation of the secretory response to nicotine.

VII. THE SECRETORY RESPONSE TO SUSTAINED DEPOLARIZATION WITH HIGH POTASSIUM CONCENTRATIONS

A. Role of Ca Channels

Baker and Rink[52] studied in detail the kinetics of the secretory response to high K in the bovine adrenal medulla. They noted that the decline of secretion during prolonged treatment with high K and Ca roughly paralleled the decline of responsivity when Ca was added at various times after the start of high K treatment and concluded that the transient character of the high K response was primarily due to inactivation of the voltage-dependent Ca channel. These findings paralleled the observation that in the isolated squid axon, depolarization-evoked Ca influx, as monitored by equorin fluorescence, rises to a peak in a few seconds and falls to a new steady level in 0.5 to 5 min.[53] It similarly appears that inactivation of voltage-sensitive Ca permeability may contribute to the decline of K-evoked secretion in the cat adrenal gland, since late Ca addition reduces the secretory response to high K in this preparation.[49,54]

Ca channel inactivation may be due to two different processes, both located on the gating mechanisms of the Ca channel: (1) changes on kinetics of openings and closings during the sustained stimulus, and (2) probability of Ca channel reopenings once closed to an inactivated state.

The first alternative seems unlikely, because BAY-K-8644, which is known to affect these kinetics,[40,41] did not modify at all the rates of inactivation of catecholamine release (Figure 2C and D). The second one might be equivalent to the process called ''run down'' or ''wash-out'' from electrophysiological studies and might suggest a decrease in the number of available channels. This rundown process was studied on chromaffin cell Ca channels

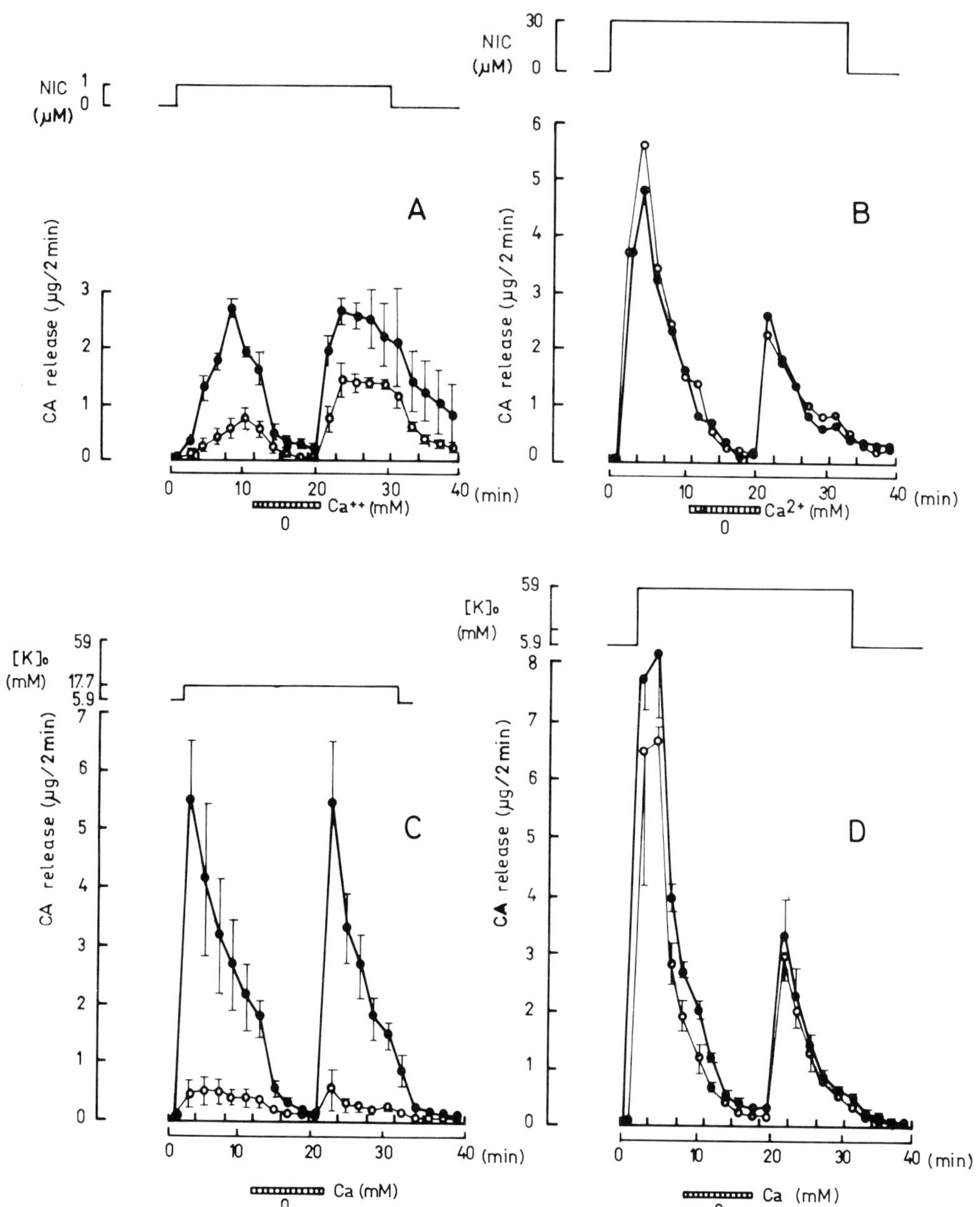

FIGURE 2.   Catecholamine (CA) release evoked by 1 (A) and 30 (B) μM nicotine or 17.7 (C) and 59 (D) mM K from cat adrenal glands perfused with oxygenated Krebs-bicarbonate solution at room temperature. These figures show the effect of Ca_o removal and its latter reintroduction (2.5 mM, bottom bar) always in the presence of the secretagogue (top horizontal line). In all the experiments one gland was used as control (○), and the contralateral one was perfused with 1 μM BAY-K-8644 (●). Data are means ± S.E.M. of four paired experiments.

by Fenwick et al.[32] and seems to be accelerated by higher potential stimulation and by higher Ca_i concentrations. The conclusions of recent studies in a number of excitable cells, including Paramecium[34] and the neuron of Aplysia,[35] suggest that the voltage-dependent Ca permeability may be inactivated by a Ca-dependent mechanism (for review, see Eckert and Chad[46]).

If it is true that Ca plays a role in controlling the rate of inactivation of the secretory response, some divalent cations other than Ca might exhibit different rates of inactivation.

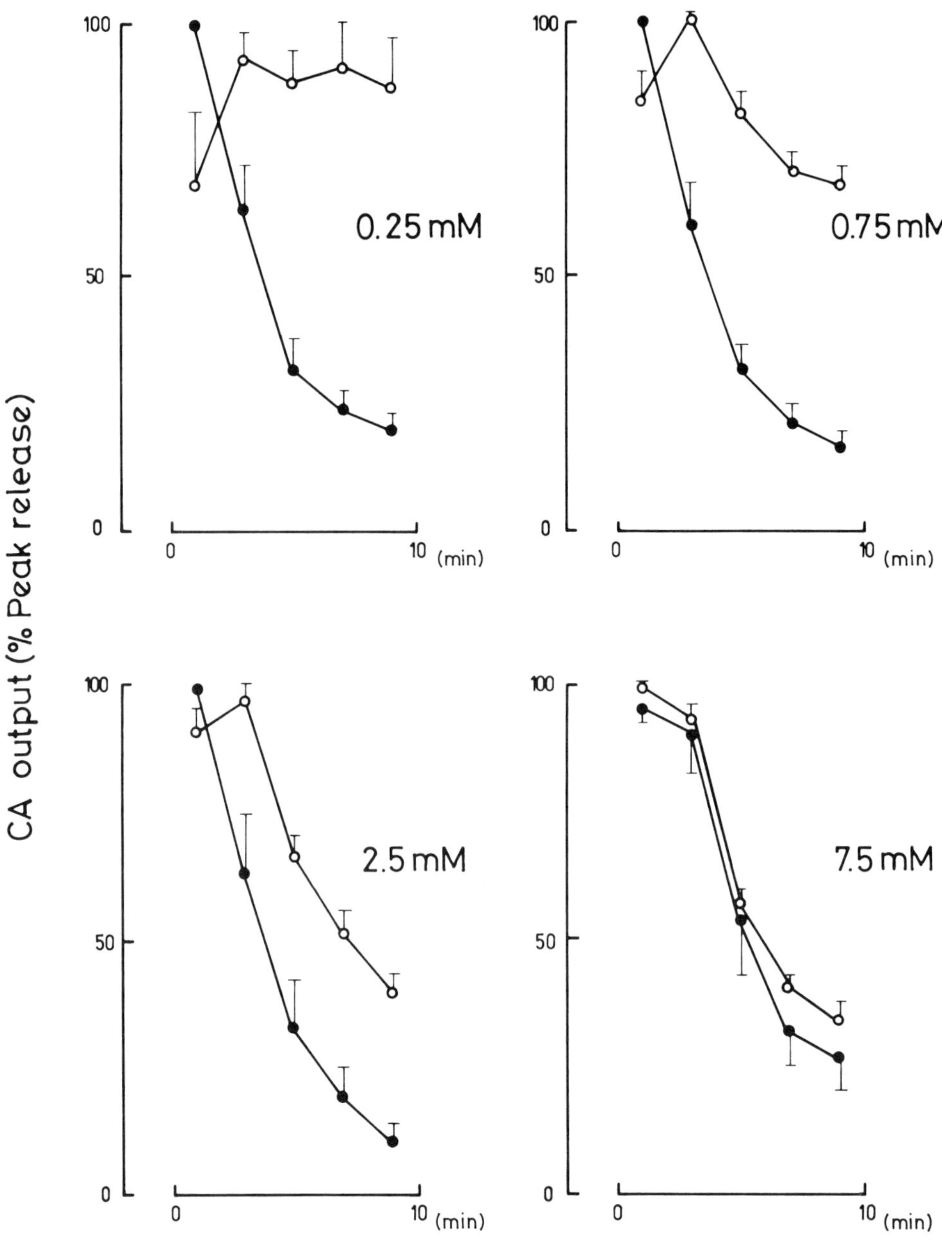

FIGURE 3.   Rates of inactivation of the catecholamine (CA) secretory responses evoked by 10-min pulses of 59 mM K at several concentrations of Ca^{2+} (●) or Sr^{2+} (○). Cat adrenal glands were perfused at room temperature with oxygenated Krebs-bicarbonate solution and samples collected at 2-min intervals; CA outputs are expressed as % of the peak releases obtained in each high K stimulation. Data are means ± S.E. of three paired experiments.

Since Sr and Ba can substitute for Ca in releasing catecholamines from cat adrenal glands,[42] it was interesting to test their ability to inactivate the secretory response.

Figure 3 shows a surprising finding. Inactivation of the secretory response to 59 mM K is not seen when using 0.25 mM Sr instead of Ca; at 0.75 and 2.5 mM concentrations the rates of decline are much slower in Sr than in Ca and at 7.5 mM both are similar. With Ba, the secretory effects of high K did not inactivate at all (data not shown). Nevertheless, the different inactivation rates observed in the presence of Ca, Sr, or Ba could be partly explained on the grounds of their relative affinities for an intracellular site located on the secretory machinery and controlling the rates of exocytosis. Little is known concerning the relative affinities of these three cations on intracellular proteins related to the stimulus-secretion coupling mechanism. A recent report[55] demonstrates that Ca, Sr, and Ba stimulate equally both exocytosis and polyphosphoinositide hydrolysis in the sea urchin egg. Mellow[56] concluded also that at the frog muscle endplate there is divalent cation nonselectivity at intra-terminal releasing sites. However, neither work explores the rates of inactivation of such secretory responses.

A second site where the three cations might behave differently is at intracellular organelles that sequester Ca. It is known that mitochondria and the sarcoplasmic reticulum in many cell types take up Ca and Sr, but not Ba, at significant rates.[57] There is, however, no information about the selectivity of chromaffin cell organelles for different divalent cations; work along this line is proceeding in our laboratory.

With the data presently available it seems difficult to establish a firm correlation between Ca channel inactivation and the declining rate of the secretory response. On the other hand, recent results using Quin-2 to monitor Ca_i levels[58] during sustained depolarizations are contradictory. Knight and Kesteven[33] demonstrated that the rates of increase and decrease of both Ca_i and catecholamine release closely paralleled. Their experiments indicate that a relationship between the activity of the Ca channel and inactivation of the secretory response might exist. However, more recently, long-lasting persistent increases of Ca_i using the same system on PC12 cells[59] or chromaffin cells[60] have been reported, suggesting that Ca channel inactivation is not involved in the decline of the secretory response.

B. Are the Constant Levels of Intracellular Calcium Responsible for the Decline of Secretion?

As suggested by Schiavone and Kirpekar,[49] the maintenance of Ca_i levels within a plateau might account for the transient nature of the secretory response to high K. This suggestion is supported by experiments reported by Knight and Baker using leaky chromaffin cells stimulated with several Ca concentrations.[61] A decline of the secretory rate was observed with a kinetic rather similar to that obtained in intact cells; the later addition of more Ca reactivated the secretory response. This was also seen at the perfused cat adrenal.[49] Then, the secretory effect of high K may decline, in part, because the intracellular stimulus-secretion coupling mechanism may become refractory to activation, while Ca_i is maintained constant and is not increasing. Amy and Kirshner[62] have reported that phosphorylation of two cellular protein fractions accompanies Ca-dependent activation of catecholamine secretion by nicotine, veratridine, ionomycin, or Ba in the bovine adrenal medulla, and Baker and Knight[61] proposed that phosphorylation of some intracellular proteins may play a role in the decline of secretion evoked by Ca addition to a preparation of leaky chromaffin cells. According to the proposals of such authors, the rate of catecholamine secretion would be proportional to dCa_i/dt. The following recent experiments from our laboratory agree with this assumption.

In Figure 2C and D, it can be seen that prolonged depolarization of perfused glands in the presence of constant concentrations of extracellular Ca invariably leads to inactivation, almost to basal levels, of the catecholamine secretory response. Provided that both raised K and Ca concentrations remain unchanged, the release process will be kept desensitized. If constant levels of Ca_i are involved in such inactivation process, it is likely that Ca removal from the perfusion solution (always in the presence of high K to keep chromaffin cells continuously depolarized) could cause a decrease of intracellular-free Ca concentrations and reactivate the secretory mechanism upon Ca reintroduction. This was the case. In experiments shown in Figure 2C, both glands were perfused with 17.7 m*M* K for 30 min in the presence or the absence of BAY-K-8644 (1 µ*M*); the drug produced its typical potentiation of the secretory response. Once the release process desensitized, Ca was removed for a 10-min period; its reintroduction caused the full restoration of the secretory response, with peak release and time-course of activation and inactivation closely similar to the response obtained initially upon increase of K from 5.9 to 17.7 mM. In fact, the second response looks like a mirror image of the first response. The typical enhancement of secretion was observed in the presence of BAY-K-8644.

If Ca entry and its subsequent intracellular accumulation are responsible for the desensitization of catecholamine release, then interruption of both processes by the inorganic Ca entry blocker Co should lead to results similar to those obtained with the Ca omission experiments described above. This was the case. The addition of Co (5 mM) once the high K-evoked secretory response inactivated caused the full reactivation of the secretory response upon Co removal from the perfusion medium (data not shown).

At higher depolarizations (59 mM K), reactivation of the secretory response upon Ca removal and reintroduction, or Co addition and removal, was only partial (Figure 2D), speaking in favor of an irreversible voltage-dependent component in the process of inactivation of catecholamine release. The fact that the secretory response is recovered (even at low rates of secretion, in 0.25 mM Ca) only by 40% when a strong depolarizing stimulus is used indicates that the inactivation is also linked to irreversible Ca channel inactivation. Another way to test this hypothesis was with experiments in which adrenomedullary cells of perfused cat adrenal glands were maintained continuously depolarized with a 59-mM K solution, while the Ca concentration was being increased from 0 to 1, 5, or 10 mM by using a continuous concentration gradient system, in contrast to the protocols used by Knight and Baker[61] and Schiavone and Kirpekar[49] that increased Ca$_o$ in a step-wise manner. A peak secretory response was reached soon after Ca started entering the gland (Figure 4). The surprising observation was that the secretory response did not decline as long as Ca was being raised. Maintenance of the secretory response at a plateau was always seen with gradients in a wide range of Ca concentrations (0 to 1, 0 to 5, 0 to 10 mM), suggesting that a critical intracellular level of Ca was not reached in this situation. However, in the presence of BAY-K-8644 (1 μM), the secretory response was potentiated in all Ca gradients; the release process was not inactivated in the 0- to 1-mM Ca gradient, but in the gradients 0 to 5 or 0 to 10 mM the secretory response declined quickly when the gradient reached the Ca concentration around 2.5 mM.

These latter experiments allow the establishment of two tentative conclusions: (1) that the catecholamine release response to sustained high K depolarization does not inactivate provided that the external Ca concentration (which is the immediate source of Ca required at intracellular sites to trigger the exocytotic event) is increased continuously; and (2) that inactivation may take place, in spite of continued Ca increments, when the Ca channel activator BAY-K-8644 promotes a massive Ca entry that allows Ca$_i$ to reach a critical level to cause inactivation of the secretory response. In contrast to the simultaneous application of K at fixed Ca concentrations, where inactivation takes place similarly at all Ca concentrations tested, here we have found an experimental manipulation showing that in different Ca concentrations the rate of activation and inactivation of the secretory response is strongly Ca dependent.

To pursue further the idea that as Ca$_i$ reaches a plateau the secretory response inactivates, a two-step Ca gradient was performed (Figure 5). In the first step Ca was raised from 0 to 1 mM in a K-depolarized gland; the secretory response was increasing slowly in both glands until reaching the usual plateau of secretion. When the gradient was stopped in one of the two glands and perfused with the fixed concentration of 1 mM Ca, the secretory response started to decline immediately. Upon reinitiation of the Ca gradient from 1 to 2 mM, the declining tendency of the secretory response stopped and a stable noninactivating secretory response was again obtained. The contralateral gland perfused with a single-step Ca gradient (0 to 2 mM) gave a secretory profile with the usual plateau; again, inactivation of the

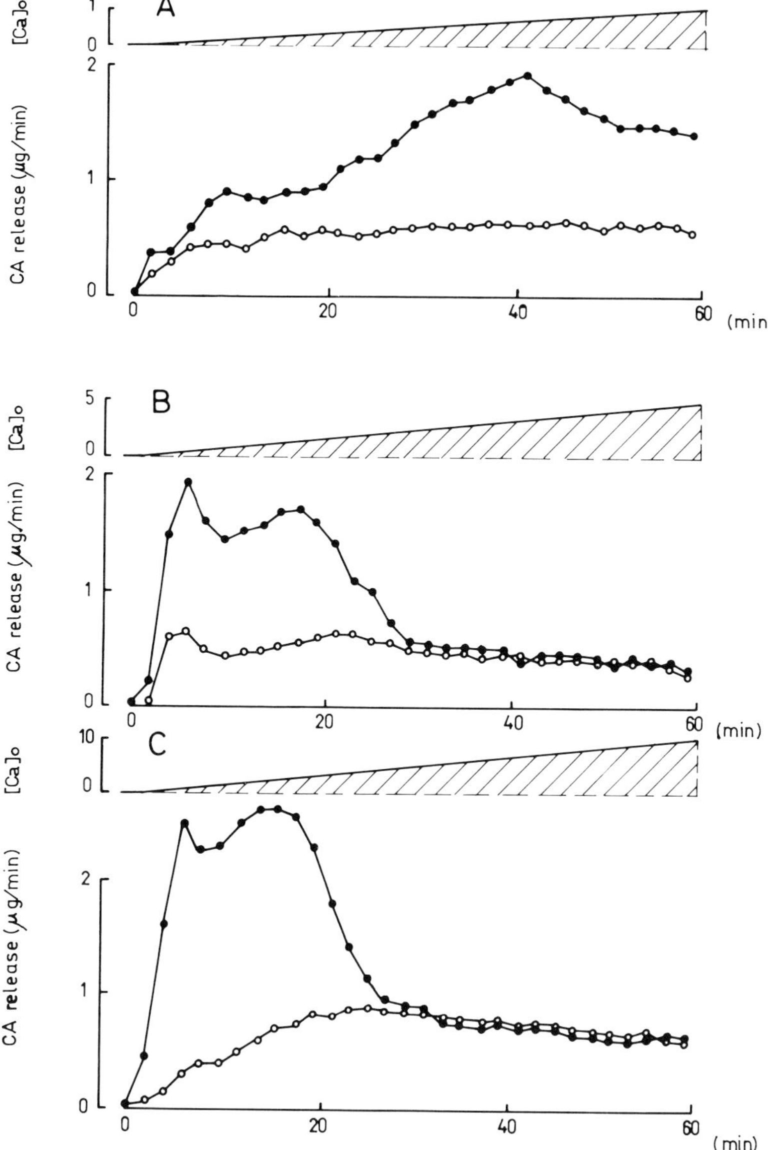

FIGURE 4.   Inactivation of the catecholamine (CA) secretory response evoked by high K at the perfused cat adrenal gland is prevented by continuously raising the extracellular Ca concentration. The secretory stimulus was 59 mM K during 60 min in all the experiments (A, B, and C). The ranges of the Ca$_o$ gradients were from 0 to 1 mM (A), from 0 to 5 mM (B), and from 0 to 10 mM (C); (○) control glands, (●) BAY-K-8644-treated glands (1 μM).

secretory response was initiated when the gradient tended and then a fixed Ca concentration (2 mM) was perfused (Figure 5B). In this experiment, it is curious to note the profile of the secretory response to 59 mM K obtained at the end of the experiment in the presence of a fixed 2.5-mM Ca concentration. Observe the sharp contrast between the rates of activation of the responses obtained by adding high K plus a fixed concentration of Ca simultaneously, with those obtained at the beginning of the experiment with the Ca gradients, and especially, the fast inactivation rate observed when depolarization is applied to a gland perfused previously with a fixed Ca concentration, in contrast to the lack of inactivation observed with the Ca gradient.

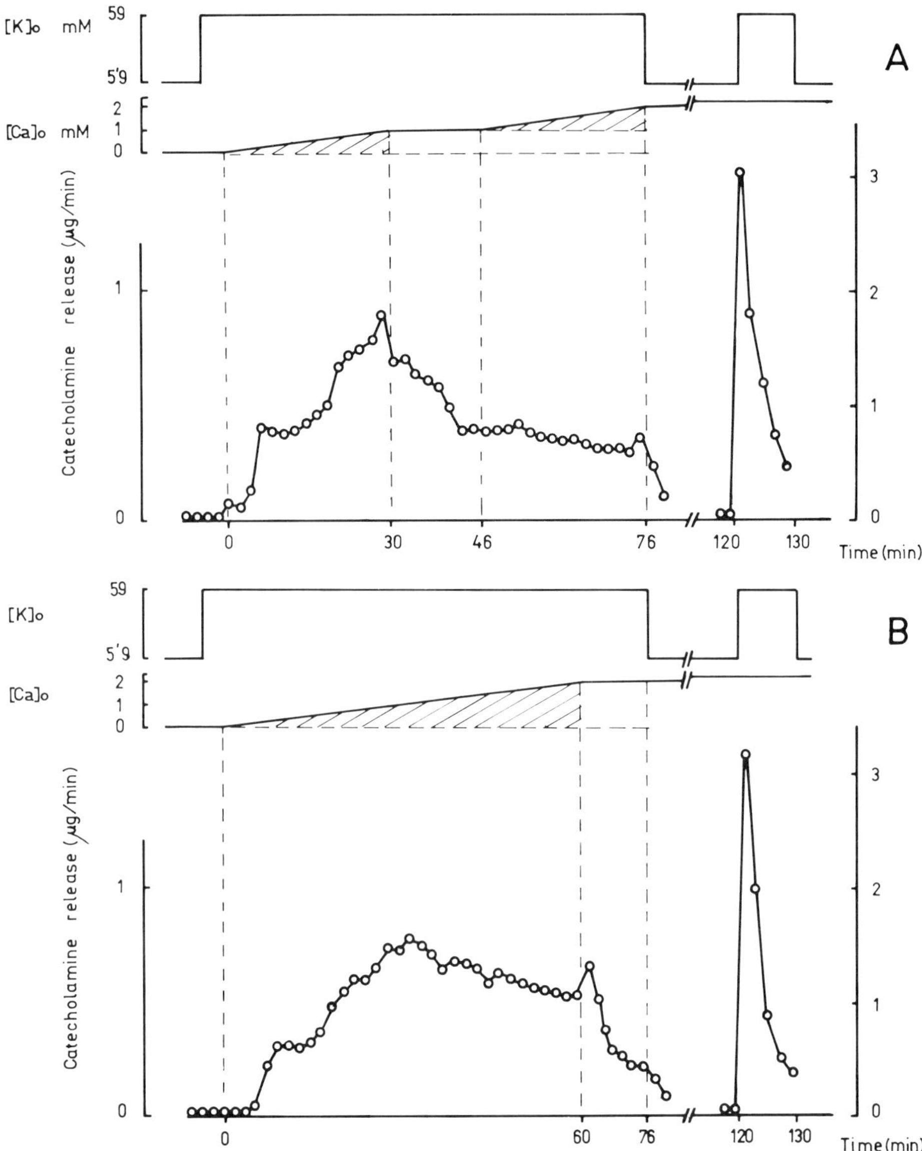

FIGURE 5.   Catecholamine release evoked by high K (59 m*M*) in the presence of a Ca_o gradient. Gland A was perfused by using a $(Ca)_o$ gradient from 0 to 1 m*M* during the first 30 min. After that, the $(Ca)_o$ was fixed at 1 m*M* during the following 16 min, and it was increased again as a gradient from 1 to 2 m*M* during another 30-min period. Gland B was perfused by using a Ca_o gradient from 0 to 2 m*M* during the first 60 min. The final concentration was kept constant at 2 m*M* during an additional period of 16 min in depolarizing conditions. At the end of the experiment a 10-min pulse of high K was given to both glands in order to rule out catecholamine stores depletion and to verify their functional state. Data are from a single paired experiment.

Since in the presence of Sr the K-evoked response inactivated very slowly (Figure 3), it was also interesting to observe its behavior in glands depolarized with 59 m*M* K and giving Sr gradually as a rising continuous concentration gradient in protocols similar to those performed with Ca. In Figure 6 it can be appreciated that Sr can substitute for Ca very efficiently and that it evoked a greater secretory response.

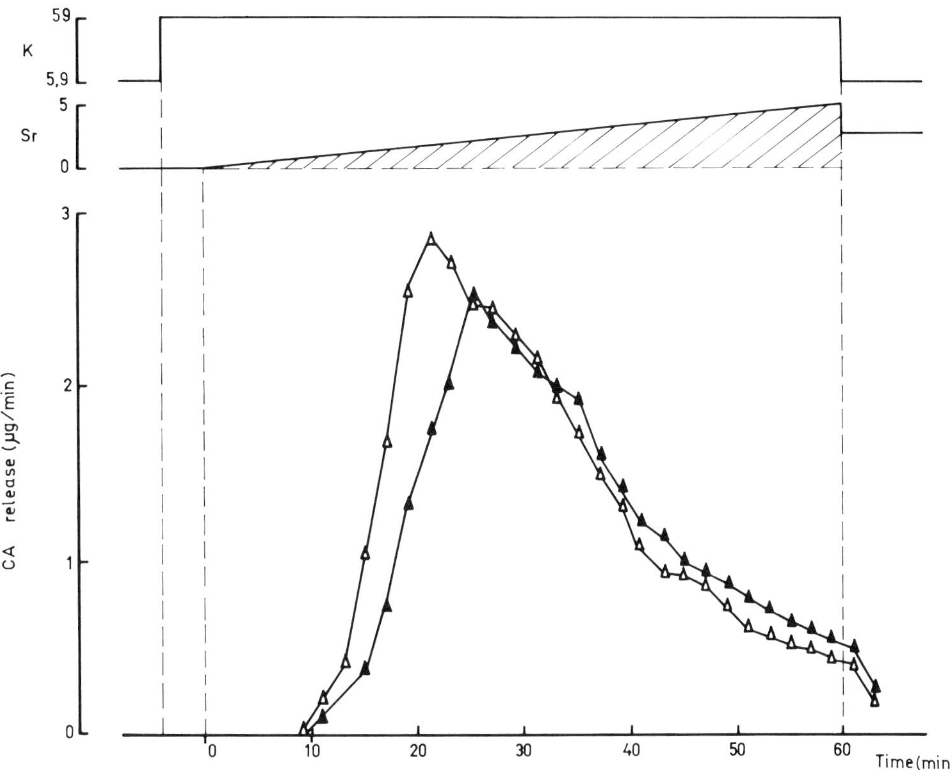

FIGURE 6.   Catecholamine release evoked by high K (59 mM) in the presence of a Sr_o gradient from 0 to 5 mM. All solutions contained 1 mM EGTA. (▲) control; (△) BAY-K-8644, 1 μM. Data are from a single paired experiment.

Similar secretory profiles were obtained when using Ca gradients + BAY-K-8644 or Sr gradients in the absence of BAY-K-8644 (i.e., plateau followed by fall to another plateau similar to that obtained when using Ca gradient in the absence of BAY-K-8644, Figure 4B and C); again, these data strongly suggest that the differences observed when comparing Ca and Sr gradients are located on the target for the drug. In other words, there might be a specific site with affinity for divalent cations associated to Ca channels of the chromaffin cells, and such an affinity is modified by the dihydropyridine. As higher the affinity of the divalent cation is on the Ca channel, greater will be the extent of the modifications evoked by BAY-K-8644 on the rates of catecholamine release and vice versa.

VIII. CONCLUSIONS AND PERSPECTIVES

A tentative model to explain how the catecholamine release cycle at the chromaffin cell might be modulated by voltage-sensitive Ca channel activity and by Ca_i levels has been suggested by studying the mechanism of action of the novel Ca channel activator BAY-K-8644. Thus, the secretory response can be explained in the following terms (Figure 1).

A depolarizing stimulus activates Ca channels and promotes a fast rise of Ca_i concentrations. The level of Ca_i reached by a given stimulus depends on the size of the stimulus and on the outside-inside Ca concentration gradient. The initial rate of catecholamines released (peak response) also depends on the size of the stimulus and on the outside-inside Ca concentration gradient.[7] In other words, the rate of secretion is directly proportional to dCa_i/dt, and its extent and duration depends on the final Ca_i concentration achieved.[61]

If the activity of the Ca channel is controlled by Ca_i levels, it is likely to assume that a steady state will be reached early after the application of the stimulus, and thereafter Ca_i concentrations will be kept constant.[60] The decrease in this level reported by some authors[33] may be due to irreversible closing of Ca channels probably associated with the rundown process.[32] This decrease is not incompatible with the above assumption, since both situations (fixed and decreasing levels) will lead to inactivation of the secretory response. The Sr or Ba inability to inactivate catecholamine release may be due either to smaller affinity for an intracellular secretory machinery inactivation site or to their inability to promote the Ca channel rundown process.

BAY-K-8644 enhances secretory responses by promoting longer-lasting openings of the Ca channel that will allow higher Ca_i concentrations. The assumption that its effects are Ca dependent is supported by the following findings: (1) the lack of potentiation when Ca_i levels are presumably saturating (see figures of Ca gradients); (2) the similarity of the secretory curve profiles obtained with Sr gradients alone or Ca gradients plus BAY-K-8644; and (3) its smaller or absent effects on K-evoked secretion in the presence of Sr or Ba, respectively.

There is little doubt that Ca plays a central role, not only in the triggering of the physiological catecholamine release response at the adrenal medulla, but also in limiting the extent and the time course of the secretory profile. The control of Ca metabolism and the maintenance of Ca_i levels are complex phenomena that depend on membrane ionic transients (Ca entry and Ca efflux through channels and carriers) and upon cytosolic events (Ca sequestering systems). To correlate Ca_i levels, membrane Ca transients or Ca channels activity with the kinetics of activation and inactivation of the secretory response, and simultaneous monitoring of these parameters in the same cell are required. This is not technically possible at the moment, unless a "giant" chromaffin cell is available. However, different technical approaches using isolated cultured chromaffin cells are facilitating, and will facilitate further our understanding of the modulation by Ca of the kinetics of the chromaffin cell secretory response. Such multidisciplinary approach to this exciting question is summarized in the scheme of Figure 7.

ACKNOWLEDGMENTS

We thank Prof. F. Hoffmeister (BAYER AG) for the generous supply of BAY-K-8644. Experiments reported here from our laboratory have been supported by grants from C.A.I.C.Y.T. (Ministerio de Educación y Ciencia), F.I.S.S. (Ministerio de Sanidad y Consumo), and Fundación Ramón Areces, Spain. We thank Mrs. Natividad Tera for typing the manuscript.

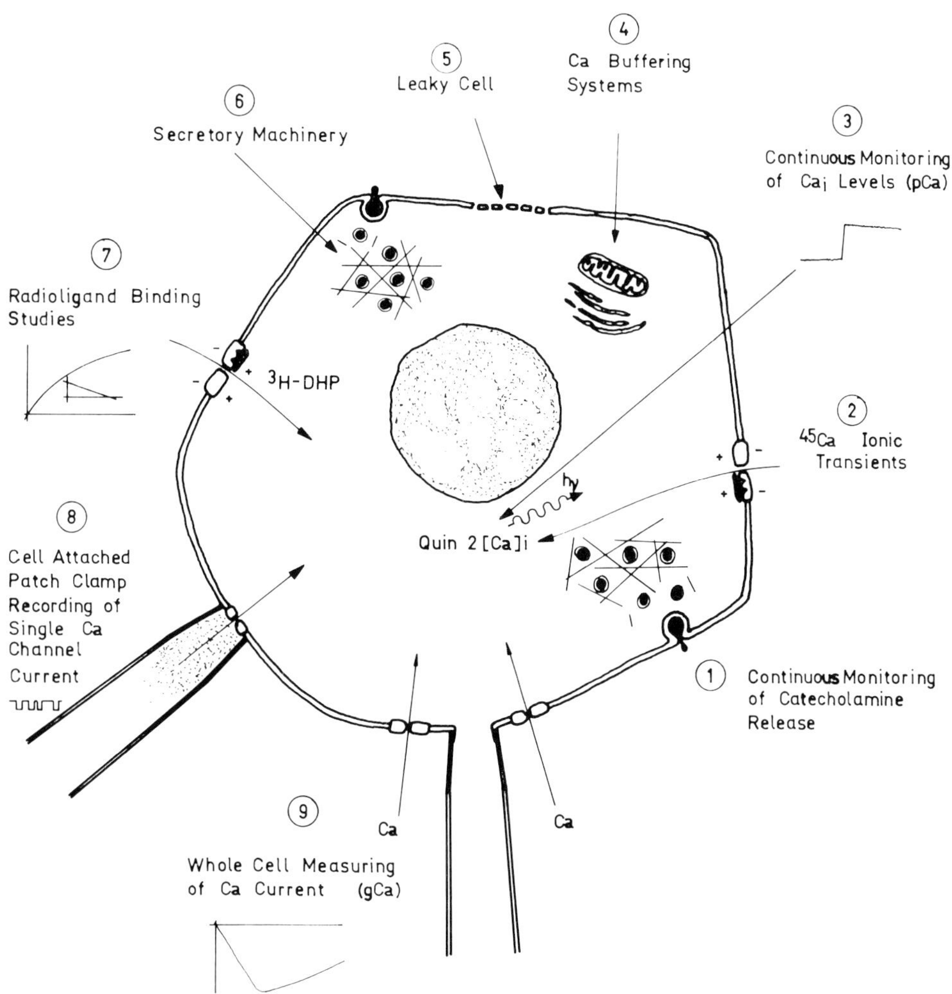

FIGURE 7.   Scheme showing a chromaffin cell and the possible technical approaches (numbered 1 to 9) that can be used to try to answer the question of how the metabolism of Ca ions correlates with the kinetics of activation and inactivation of the catecholamine secretory response.

REFERENCES

1. **Feldberg, W., Minz, B., and Tsudzimura, H.,** The mechanism of the nervous discharge of adrenaline, *J. Physiol. (London)*, 81, 286, 1934.
2. **Collier, B., Johnson, G., Kirpekar, S. M., and Prat, J.,** The release of acetylcholine and of catecholamine from the cat's adrenal gland, *Neuroscience*, 13, 957, 1984.
3. **Douglas, W. W. and Rubin, R. P.,** The role of calcium in the secretory response of the adrenal medulla to acetylcholine, *J. Physiol. (London)*, 159, 40, 1961.
4. **Ceña, V., Nicolás, G. P., Sánchez-García, P., Kirpekar, S. M., and García, A. G.,** Pharmacological dissection of receptor-associated and voltage-sensitive ionic channels involved in catecholamine release, *Neuroscience,* 10, 1455, 1983.
5. **García, A. G., Ceña, V., and Frias, J.,** Pharmacological dissection of ionic channels involved in catecholamine release from the chromaffin cell, in *Actualités Chimie Therapeutique,* 11th Series, Lavoisier, Paris, 1984, 165.

6. **García, A. G., Sala, F., Ladona, M. G., Ceña, V., and Montiel, C.,** Analysis of the catecholamine secretory process by using a novel dihydropyridine calcium "agonist" and potassium or calcium gradients, in *Regulation of Transmitter Function,* Vizi, E. S. and Magyar, K., Eds., Elsevier, Amsterdam, 1984.

7. **Ishikawa, K. and Kanno, T.,** Influences of extracellular calcium and potassium concentrations on adrenaline release and membrane potential in the perfused adrenal medulla of the rat, *Jpn. J. Physiol.,* 28, 275, 1978.

8. **Holz, R. W., Senter, R. A., and Frye, R. A.,** Relationship between Ca^{2+} uptake and catecholamine secretion in primary dissociated cultures of adrenal medulla, *J. Neurochem.,* 39, 635, 1982.

9. **Kilpatrick, D. L., Slepetis, R. J., Corcoran, J. J., and Kirshner, N.,** Calcium uptake and catecholamine secretion by cultured bovine adrenal medulla cells, *J. Neurochem.,* 38, 427, 1982.

10. **Douglas, W. W. and Rubin, R. P.,** The mechanism of catecholamine release from the adrenal medulla and the role of calcium in stimulus-secretion coupling, *J. Physiol. (London),* 167, 288, 1963.

11. **Douglas, W. W. and Poisner, A. M.,** Preferential release of adrenaline from the adrenal medulla by muscarine and pilocarpine, *Nature (London),* 208, 1102, 1965.

12. **Kirpekar, S. M., Prat, J. C., and Schiavone, M. T.,** Effect of muscarine on release of catecholamines from the perfused adrenal gland of the cat, *Br. J. Pharmacol.,* 77, 455, 1982.

13. **Fleckenstein, A.,** Specific pharmacology of calcium in myocardium, cardiac pacemakers, and vascular smooth muscle, *Ann. Rev. Pharmacol. Toxicol.,* 17, 149, 1977.

14. **Spedding, M.,** Calcium antagonist subgroups, *TIPS,* 6, 109, 1985.

15. **Fleckenstein, A.,** Die bedentung der energiereichen phosphosphate für kontraktilität and tonus des myokards, *Verh. Dtsch. Ges. Inn. Med.,* 70, 81, 1964.

16. **Pinto, J. E. B. and Trifaró, J. M.,** The different effects of D-600 (methoxy-verapamil) on the release of adrenal catecholamines induced by acetylcholine, high potassium or sodium deprivation, *Br. J. Pharmacol.,* 57, 127, 1976.

17. **Lemaire, S., Derome, G., Tseng, R., Mercier, P., and Lemaire, J.,** Distinct regulations by calcium of cyclic GMP levels and catecholamine secretion in isolated bovine adrenal chromaffin cells, *Metabolism,* 30, 462, 1981.

18. **Wakade, A. R.,** Studies on secretion of catecholamines evoked by acetylcholine or transmural stimulation of the rat adrenal gland, *J. Physiol. (London),* 313, 463, 1981.

19. **Norris, P. J., Dhaliwal, D. K., Druce, D. P., and Bradford, H. F.,** The suppresion of stimulus-evoked release of amino acid neurotransmitters from synaptosomes by verapamil, *J. Neurochem.,* 40, 514, 1983.

20. **El-Fakahany, E. and Richelson, E.,** Effect of some calcium antagonists on muscarinic receptor-mediated cyclic GMP formation, *J. Neurochem.,* 40. 705, 1983.

21. **Galzin, A. M. and Langer, S. Z.,** Presynaptic alpha$_2$-adrenoceptor antagonism by verapamil but not by diltiazem in rabbit hypothalamic slices, *Br. J. Pharmacol.,* 78, 571, 1983.

22. **Bregestovski, P. D., Miledi, R., and Parker, I.,** Calcium conductance of acetylcholine-induced endplate channels, *Nature (London),* 279, 638, 1979.

23. **Takahashi, M. and Ogura, A.,** Dihydropyridines as potent calcium channel blockers in neuronal cells, *FEBS Lett.,* 152, 191, 1983.

24. **Douglas, W. W. and Poisner, A. M.,** On the mode of action of acetylcholine in evoking adrenal medullary secretion: increased uptake of calcium during the secretory response, *J. Physiol. (London),* 162, 385, 1962.

25. **García, A. G., Sala, F., Reig, J. A., Viniegra, S., Frias, J., Fonteriz, R., and Gandia, L.,** Dihydropyridine BAY-K-8644 activates chromaffin cell calcium channels, *Nature (London),* 309, 67, 1984.

26. **Baker, P. F. and Knight, D. E.,** Calcium control of exocytosis in bovine adrenal medullary cells, *TINS,* 7, 120, 1984.

27. **Douglas, W. W., Kanno, T., and Sampson, S. R.,** Influence of the ionic environment on the membrane potential of adrenal chromaffin cells and on the depolarizing effect of acetylcholine, *J. Physiol. (London),* 191, 107, 1967.

28. **Douglas, W. W. and Kanno, T.,** The effect of amethocaine on acetylcholine induced depolarization and catecholamine secretion in the adrenal chromaffin cell, *Br. J. Pharmacol.,* 30, 612, 1967.

29. **Biales, B., Dichter, M., and Tischler, A.,** Electrical excitability of cultured adrenal chromaffin cells, *J. Physiol. (London),* 262, 743, 1976.

30. **Brandt, B. L., Hagiwara, S., Kodokoro, Y., and Miyazaki, S.,** Action potentials in the rat chromaffin cells and effects of acetylcholine, *J. Physiol. (London),* 263, 417, 1976.

31. **Kidokoro, Y., Miyazaki, S., and Ozawa, S.,** Acetylcholine-induced membrane depolarization and potential fluctuations in the rat adrenal chromaffin cell, *J. Physiol. (London),* 324, 203, 1982.

32. **Fenwick, E. M., Marty, A., and Neher, E.,** Sodium and calcium channels in bovine chromaffin cells, *J. Physiol. (London),* 331, 599, 1982.

33. **Knight, D. E. and Kesteven, N. T.,** Evoked transient intracellular free Ca^{2+} changes and secretion in isolated bovine adrenal medullary cells, *Proc. R. Soc. London Ser. B,* 218, 177, 1983.

34. **Brehm, P. and Eckert, R.,** Calcium entry leads to inactivation of calcium channel in paramecium, *Science,* 202, 1203, 1978.

35. **Tillotson, D.**, Inactivation of Ca conductance dependent on entry of Ca ions in molluscan neurons, *Proc. Natl. Acad. Sci. U.S.A.*, 76, 1497, 1979.

36. **Schramm, M., Thomas, G., Towart, R., and Franckowiak, G.**, Novel dihydropyridines with positive inotropic action through activation of Ca^{2+} channels, *Nature (London)*, 303, 535, 1983.

37. **García, A. G., Kirpekar, S. M., and Prat, J. C.**, A calcium ionophore stimulating the secretion of catecholamines from the cat adrenal, *J. Physiol. (London)*, 244, 253, 1975.

38. **Carvalho, M. H., Prat, J. C., García, A. G., and Kirpekar, S. M.**, Ionomycin stimulates secretion of catecholamines from cat adrenal gland and spleen, *Am. J. Physiol.*, 242, E137, 1982.

39. **Montiel, C., Artalejo, A. R., and García, A. G.**, Effects of the novel dihydropyridine BAY-K-8644 on adrenomedullary catecholamine release evoked by calcium reintroduction, *Biochem. Biophys. Res. Commun.*, 120, 851, 1984.

40. **Hess, P., Lansman, J. B., and Tsien, R.**, Different modes of Ca channel gating behaviour favoured by dihydropyridine Ca agonists and antagonists, *Nature (London)*, 311, 538, 1984.

41. **Koubun, S. and Reuter, H.**, Dihydropyridine derivates prolong the open state of Ca channels in cultured cardiac cells, *Proc. Natl. Acad. Sci. U.S.A.*, 81, 4824, 1984.

41a. **Artulejo, C. R. and García, A. G.**, Effects of BAY-K-8644 on cat adrenal catecholamine secretory responses to A23187 or ouabain, *Br. J. Pharmacol.*, 88, 757, 1986.

42. **Douglas, W. W. and Rubin, R. P.**, The effects of alkaline earths and other divalent cations on adrenal medullary secretion, *J. Physiol. (London)*, 175, 231, 1964.

43. **Hagiwara, S. and Byerly, L.**, Calcium channel, *Ann. Rev. Neurosci.*, 4, 69, 1981.

44. **Tsien, R. W.**, Calcium channels in excitable cell membranes, *Ann. Rev. Physiol.*, 45, 341, 1983.

45. **Nachshen, D. A. and Blaustein, M. P.**, Influx of calcium, strontium, and barium in presynaptic nerve endings, *J. Gen. Physiol.*, 79, 1065, 1984.

46. **Eckert, R. and Chad, J. E.**, Inactivation of Ca channels, *Prog. Biophys. Mol. Biol.*, 44, 215, 1984.

47. **Hoshi, T., Rothlein, J., and Smith, S. J.**, Facilitation of Ca^{2+}-channel currents in bovine adrenal chromaffin cells, *Proc. Natl. Acad. Sci. U.S.A.*, 81, 5871, 1984.

48. **Brown, A. M., Kunze, D. L., and Yatani, A.**, The agonist effect of dihydropyridines on Ca channels, *Nature (London)*, 311, 570, 1984.

49. **Schiavone, M. T. and Kirpekar, S. M.**, Inactivation of secretory responses to potassium and nicotine in the cat adrenal medulla, *J. Pharmacol. Exp. Ther.*, 223, 743, 1982.

50. **Manthey, A. A.**, The effect of calcium on the desensitization of membrane receptors at the neuromuscular junction, *J. Gen. Physiol.*, 49, 963, 1966.

51. **Nastuk, W. L.**, Cholinergic receptors desensitization, in Synapses, Cottrell, G. A. and Usherwood, P. N. R., Eds., Academic Press, New York, 1977, 177.

52. **Baker, P. F. and Rink, T. J.**, Catecholamine release from bovine adrenal medulla in response to maintained depolarization, *J. Physiol. (London)*, 253, 593, 1975.

53. **Baker, P. F., Meves, H., and Ridgway, E. B.**, Calcium entry in response to maintained depolarization of squid axons, *J. Physiol. (London)*, 231, 527, 1973.

54. **Kirpekar, S. M., García, A. G., and Schiavone, M. T.**, Secretion of catecholamines from the adrenal gland by various agents, in *Advances in the Biosciences*, Vol. 36, Izumi, F., Oka, M., and Kumakura, K., Eds., Pergamon Press, Oxford, 1982, 55.

55. **Whitaker, M. and Aitchison, M.**, Calcium-dependent polyphospoinositide hydrolysis is associated with exocytosis in vitro, *FEBS Lett.*, 182, 119, 1985.

56. **Mellow, A. M.**, Equivalence of Ca^{2+} and Sr^{2+} in transmitter release from K^+-depolarised nerve terminals, *Nature (London)*, 282, 84, 1979.

57. **Lehninger, A. L.**, Mitochondria and calcium ion transport, *Biochem. J.*, 119, 129, 1970.

58. **Tsien, R. Y., Pozzan, T., and Rink, T. J.**, Calcium homeostasis in intact lymphocytes: cytoplasmic free calcium monitored with a new intracellularly trapped fluorescent indicator, *J. Cell Biol.*, 94, 325, 1982.

59. **Meldolesi, J., Huttner, W. B., Tsien, R. Y., and Pozzan, T.**, Free cytoplasmic Ca^{2+} and neurotransmitter release: studies on PC12 cells and synaptosomes exposed to alpha-iatrotoxin, *Proc. Natl. Acad. Sci. U.S.A.*, 81, 620, 1984.

60. **Burgoyne, R. D. and Cheek, T. R.**, Is the transient nature of the secretory response of chromaffin cells due to inactivation of calcium channels?, *FEBS Lett.*, 182, 115, 1985.

61. **Knight, D. E. and Baker, P. F.**, Calcium-dependence of catecholamine release from bovine adrenal medullary cells after exposure to intense electric fields, *J. Membr. Biol.*, 68, 107, 1982.

62. **Amy, C. M. and Kirshner, N.**, Phosphorylation of adrenal medulla cell proteins in conjunction with stimulation of catecholamine secretion, *J. Neurochem.*, 36, 847, 1981.

Chapter 15

PEPTIDE MODULATION OF ADRENAL CHROMAFFIN CELL SECRETION

Bruce G. Livett

TABLE OF CONTENTS

I. CONTROL OF ADRENAL MEDULLARY SECRETION: THE CLASSICAL VIEW

Study of the mechanisms responsible for control of secretion from the adrenal medulla began in earnest with the observations of early physiologists that a wide range of physiological stimuli (e.g., cold, heat, emotional and physical stress, anoxia and asphyxia, pH, hypotension, insulin-induced hypoglycemia, and glucagon) act either indirectly through the splanchnic nerve or directly on the chromaffin cells themselves to increase the secretion of catecholamines into the circulation.[1] It was shown as early as 1934 by Feldberg and colleagues[2] that "the mechanism of the nervous discharge of adrenaline" from the adrenal medulla involved release of acetylcholine (ACh) from the splanchnic nerve that innervates the gland.

Progress in this area was aided greatly by concurrent studies into the chemical nature of nervous transmission at synapses and, in particular, advances in our knowledge about the nature of the nicotinic receptor at the neuromuscular junction.[3]

One characteristic of the nicotinic response seen at the neuromuscular junction in vivo and the adrenal medulla in vitro is the rapidly diminished response resulting from maintained nerve stimulation or repeated exposure to agonists, respectively. The terms "tachyphylaxis" and "desensitization" have been used to describe this phenomenon in vivo and in vitro. Given the rapid onset of desensitization when the retrogradely perfused bovine adrenal gland preparation is exposed to even moderate concentrations of agonists in vitro[4,5] (see Figure 1), it is somewhat surprising that in vivo the adrenal continues to secrete CA during a stressful encounter when ACh is being released maximally and is, presumably, flooding the nicotinic receptors on the chromaffin cells with transmitter. As a result of recent work in several laboratories, it is now appreciated that the splanchnic nerve contains, in addition to the classical neurotransmitter, ACh, a number of putative peptide neurotransmitters, and/or neuromodulators.[6] Among these, the enkephalins and substance P have been the most studied with respect to their possible modulatory effects on adrenal secretion in vitro.[7,8] It is now believed that these peptides serve a neuromodulatory role in vivo to protect the nicotinic receptor against desensitization under conditions of stress.[9] Before discussing these newer aspects of peptide control of secretion, a brief review of presently known influences on adrenal secretion is given to provide the necessary background.

A. Neuronal Control

Upon splanchnic nerve stimulation, ACh is secreted from the splanchnic nerve terminals and produces an increase in cytoplasmic calcium within the chromaffin cells that initiates the sequence of events termed "stimulus-secretion coupling" and ultimately results in catecholamine secretion. Two modes of calcium entry into chromaffin cells have been postulated:[10] one associated with the nicotinic receptor-ionophore complex, and the other associated with voltage-dependent calcium channels. The mechanism of action potential formation is thought to proceed as follows: upon ACh binding to the chromaffin cell membrane, membrane depolarization takes place with characteristic superimposing *potential fluctuations* of considerable magnitude and long duration (34 to 40 msec).[11,12] These give rise to multiple action potentials at their peaks of fluctuation. Acetylcholine-induced potential fluctuations are likely to be of physiological significance, since they would favor multiple action potentials rather than desensitization during prolonged exposure to ACh.[11]

These findings *in situ* and in vitro are supported by recent findings using preparations of isolated adrenal chromaffin cells. It has been known for some years that isolated chromaffin cells are electrically excitable and exhibit transmembrane potentials as well as agonist-induced depolarization.[13-16] In rat, gerbil, chick, and human chromaffin cells, studies using both intracellular and extracellular (suction electrode) recording techniques[17-20] have shown that the cells are spontaneously active and exhibit long-lasting overshooting action potentials

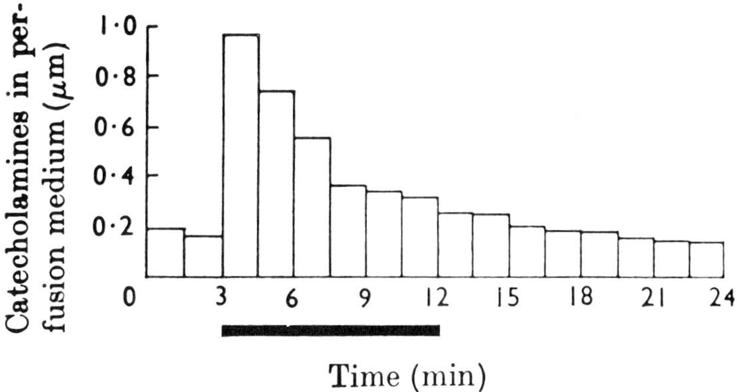

FIGURE 1.   Secretion of catecholamines by the perfused bovine adrenal gland during a prolonged infusion with carbachol. The gland was perfused through the vein with Tyrode solution, gassed with O_2 + CO_2 (95:5), and maintained at 35.5°C. The perfusion rate was approximately 6 mℓ/min. The black bar indicates a 9-min period during which 6.0 μmol of carbachol were infused per minute. (From Banks, P., *Biochem. J.*, 97, 555, 1965. With permission.)

similar to those seen in cultured sympathetic neurons. Our understanding of the membrane properties of the chromaffin cell has advanced rapidly in the last few years due to the application of the sensitive patch-clamp technique for measurement of single ion channel conductances to isolated cells.[21,22] Studies by Marty[23] on pinched-off plasma membranes isolated from bovine adrenal chromaffin cells produced by Fenwick et al.[24-26] demonstrated that these secretory cells exhibit Ca^{2+}-dependent K^+ channels with a large unitary conductance. These "big unitary conductance Ca^{2+}-dependent K^+ channels" (BK channels) are activated by depolarization at low internal Ca^{2+} levels (below 10 μM), similar to those found inside the resting chromaffin cell (unstimulated cells, $[Ca^{2+}]$ = 97 nM; ACh- or K^+-stimulated cells approximately 400 nM, as detected by Quin $2^{27,28}$). The voltage sensitivity of these channels puts them about halfway between the highly sensitive Na^+, K^+, and Ca^{2+} channels involved in action potentials, and the far less sensitive voltage-sensitive ACh-activated channels found at the vertebrate neuromuscular junction.[22] The functional significance of these BK channels remains to be determined, but it is of interest that TEA blocks the BK channels at comparatively low concentrations, in a manner similar to that reported by Kirpekar et al.,[29] for TEA inhibition of catecholamine secretion. These Ca^{2+} transients appear to generate in the multiple spiking during an action potential that may be responsible for the observed desensitization of the nicotinic receptor in the studies described above.[11,12]

In summary, the electrophysiological and patch-clamp studies of Kidokoro[11] and of Fenwick et al.[26] all point to the likely existence of mechanisms operating in vivo to protect the nicotinic secretory response against desensitization.

While early studies established that neuronal control of adrenal medullary secretion was mediated via the splanchnic nerve,[2] it was not appreciated until quite recently that the pattern of activity is an important determinant of the final secretory response. Edwards and colleagues[31] compared the effects of continuous stimulation of the peripheral ends of one or both splanchnic nerves with the effects of stimulation in bursts such that the same number of impulses were delivered per unit time, in periods ranging from 3 to 10 min. These experiments were carried out on conscious calves, employing stimuli below behavioral threshold, and involved testing neuroendocrine responses to a continuous stimulation frequency, varying between 1 and 15 Hz, with corresponding frequencies of stimulation applied for 1 sec at 10 sec-intervals (10 to 150 Hz). The results showed that the release of adrenaline (but not noradrenaline) was significantly greater in response to stimulation of the splanchnic nerve in bursts, up to and including an intermittant frequency of 40 Hz, than when continuous stimulation was applied. This is of considerable interest, because it suggests a mechanism by which these animals could alter the relative proportion of the two major catecholamines, adrenaline and noradrenaline, in response to different forms of afferent stimuli, as had been suggested by experiments in anesthetized animals.[32,33] In addition, it was found that different proportions of the two amines are secreted in young vs. older animals, and that at all ages anesthesia affects the relative amounts of two amines secreted.

These studies are of importance for several reasons: first, they sound a caution that results obtained from anesthetized animals may not pertain to the physiological situation. Second, they indicate that splanchnic nerve stimulation in vivo and ACh-induced stimulation in vitro are not necessarily equivalent in terms of the final secretory response achieved. That is to say, the splanchnic nerve contains more than just ACh capable of influencing the secretory response. As seen for salivary gland secretion where both secretory and vascular responses are controlled by the release of ACh and VIP whose ratio is critically dependent on the pattern of parasympathetic stimulation,[34,35] so, too, in the adrenal the final secretory response may depend on the effect of the pattern of stimulation on the differential release of ACh and a neuropeptide (e.g., substance P or an enkephalin). Finally, these results indicate that there is a developmental component to be considered when comparing the secretory response to a given stimulus in animals of different ages.

B. Developmental Constraints

The adrenal medulla of species such as the rat and calf is not functionally innervated by the splanchnic nerve at birth.[36] Hence, secretion from the immature adrenal is different from that in the mature animal in that it is independent of splanchnic nerve involvement. Splanchnic nerve transection,[32] nicotinic blocking agents,[37] and muscarinic blocking agents[38] are ineffective in preventing the secretory response in neonatal animals. During early fetal development, when the adrenal medulla is not yet innervated, there is a number of alternate sources of plasma catecholamines, although the quantitative contribution that each makes to the total circulating catecholamines has not been determined.[39-41] Nevertheless, it is generally held that secretion of catecholamines from the adrenal is much more important for the maintenance of homeostasis in the fetus than in the adult,[39,41-43] in which it is thought to contribute minimally.[44,45] The onset of neural control of adrenomedullary function in the rat appears at the end of the first week postnatally and matures fully by approximately 10 days of age.[36,46,47] There is, thus, a narrow window in the rat in which neurogenic and non-neurogenic responses coexist (8 to 11 days). In the neonatal and early postnatal period, non-neurogenic responses can be elicited in response to stresses such as anoxia[32] and high doses of reserpine.[46] In these immature animals[32,48] and in adult animals with adrenal denervation,[49] non-neurogenic stimuli produce an elevation in circulating catecholamines.

C. Non-Neuronal Control

Direct, non-neuronal influences on adrenal secretion have been known for some time. In 1912 Elliott[50] noted that morphine depleted the adrenal medulla of its adrenaline content, and several studies have since demonstrated both direct and neurally mediated components in the response of the adrenal medulla to both acute and chronic morphine administration. In the adult dog adrenal, morphine acts indirectly, because it fails to cause secretion of catecholamines in dogs with denervated adrenals.[51] In the mature rat, morphine and methadone evoke a reflex sympatho-adrenal discharge which results in an initial depletion of adrenal catecholamines. However, following chronic treatment, tyrosine hydroxylase and dopamine-β-hydroxylase activities increase, resulting in an elevation of catecholamine levels.[36]

The presence of high levels of enkephalins within the adrenal medulla and within the splanchnic nerve[51,52] suggests that these opioid peptides may have a role in controlling basal secretion. Some support for this idea has come from studies in vitro with cultures of bovine adrenal cells. Morphine and the opioid peptides at concentrations of 10^{-6} to 10^{-3} M have been shown to enhance the basal release of endogenous catecholamines. The enhancement by morphine (5×10^{-4} M) was Ca^{2+} dependent and stereospecific and reversed by naloxone and naltrexone.[53] It is of interest that the opioid peptides most potent in these direct actions[54] are not the pentapeptide enkephalins, but the larger opioid peptides (e.g., BAM-22P[55]). Similarly, the opioid peptides most effective at inhibiting the nicotinic receptor are the larger congeners such as BAM-22P, metorphamide,[56] and dynorphin[1-13].

A functional role for endogenous opioid peptides in damping non-neurogenic release of catecholamines has recently been proposed that may be of survival value to neonates undergoing stress (e.g., anoxia).[38] Chantry et al.[38] found that naloxone (5 mg/kg s.c.) potentiated the non-neurogenic depletion of catecholamines in the neonate induced by reserpine, while methadone (2.5 mg/kg s.c.) inhibited this non-neurogenic response. By contrast, in adult rats no potentiation of release by naloxone or inhibition by methadone was seen. Although it is not known if the levels of enkephalins in the adrenal glands are higher in the neonate than in the adult, it has been shown recently that upon chronic denervation of mature rat adrenals, a dramatic rise occurs in the level of opioid peptides.[52,57] It is also of interest that chronic treatment with reserpine leads to an increase in the synthesis of enkephalins in the adrenal medulla of adult rats.[58-60]

The secretion of adrenaline and noradrenaline from the adult adrenal medulla in response to insulin-induced stress has classically been attributed to a neurogenic mechanism.[61-64] However, there is now ample evidence that secretion of adrenal catecholamines can also occur by non-neuronal mechanisms in response to humoral factors (e.g., peptide hormones such as ACTH and glucagon) released in response to stress.[38,46,65,66] In a recent study Khalil et al.[67] showed that in response to an insulin-induced hypoglycemic stress, the adrenal secretes adrenaline by both neurogenic and non-neurogenic mechanisms. Catecholamine secretion was biphasic: (1) an early neurogenic phase lasting 30 min producing a gradual progressive increase in plasma adrenaline only and (2) a later non-neurogenic phase (40 to 60 min after insulin) during which both adrenaline and noradrenaline levels rose dramatically. The early neurogenic increase in plasma adrenaline was reduced by partial denervation and abolished by surgical complete adrenal denervation, adrenalectomy, neonatal pretreatment with capsaicin,[9] or pharmacological adrenal denervation with hexamethonium and atropine.[67] The second "non-neurogenic" phase, initiated later when the blood glucose levels fell below 75 mg%, was not altered by surgical or pharmacological adrenal denervation, showing that it was non-neuronal in origin and produced massive release of both adrenaline and noradrenaline. However, both early and late phases were abolished by adrenalectomy, showing that the adrenal was the source of both the plasma adrenaline and noradrenaline. The late response of the adrenal gland to hypoglycemia was, however, abolished by administration of glucose, showing that the non-neurogenic secretion of catecholamines is related to the level of hypoglycemia. Of significance is their observation that at 60 min after insulin (1 IU/kg) the neurogenic contribution to plasma catecholamines was only 30% of total, with the rest (70%) being contributed by non-neuronal influences. The nature of this non-neuronal influence remains to be determined.

II. RECEPTOR-MEDIATED CONTROL OF SECRETION

Studies leading to formulation of the concept of stimulus-secretion coupling in adrenal chromaffin cells[5,68] owe much to the introduction of the retrogradely perfused adrenal preparation. This simple and convenient in vitro preparation of the bovine adrenal medulla perfused retrogradely through the central lobular vein, described in 1962 by Schumann and Philippu,[69] was subsequently developed by Douglas and colleagues[68] and has been used extensively by most researchers in this field.[4,68,70,71] While the retrogradely perfused adrenal has served as an extremely valuable preparation for many studies of secretion, it is not really suitable for extensive dose-response studies of agonists and antagonists, because the secretory response of the perfused gland declines steadily, limiting the number of tests that can be carried out on one gland.

Study of postsynaptic receptor-mediated mechanisms of secretion has been facilitated greatly by the development of collagenase digestion protocols for large-scale isolation of viable and pharmacologically response adrenal chromaffin cells, together with techniques for long-term maintenance of these cells in culture.[7,24,54,72-75] Use of these adrenal cell cultures overcomes any problems associated with indirect actions of agonists and antagonists on presynaptic elements (e.g., splanchnic nerve terminals) present in the intact, retrogradely perfused adrenal. In addition, there are no endothelial diffusion barriers to limit agonist action or to delay the appearance of secreted products. Another advantage is that studies can be carried out to assess peptide interactions with postsynaptic receptor mechanisms without the complication of extracellular peptidases that complicate such studies in vivo or *in situ.*

Studies with these two preparations are in general accord and indicate that both nicotinic and muscarinic responses participate in the secretion of catecholamines from the chromaffin cell. However, the relative contribution made by nicotinic vs. muscarinic responses depends on the species being studied and on the concentration of ACh used.[75] In bovine chromaffin cells, nicotinic receptors are involved in the active secretion of catecholamines at ACh concentrations greater than 10^{-5} *M*, whereas muscarinic receptors are fully active at concentrations of 10^{-7} *M* and produce an elevation in intracellular cGMP that is inhibitory to the nicotinic response.[103]

A. Nicotinic Receptor Responses
1. The Concept of a Nicotinic Receptor-Ionophore Complex
The concept of the nicotinic receptor as a multimolecular membrane complex containing both receptor and ionophore functions comes from detailed structure-function studies and biochemical and biophysical investigations of the nicotinic binding site in other tissues. In the electroplax of the *Torpedo* electric eel, for example, the concentration of the nicotinic receptor is so high as to allow its isolation and characterization.[76] These studies reveal that the receptor is a multimolecular, integral, membrane protein complex consisting of five subunits ($\alpha_2\beta\gamma\delta$) arranged in a doughnut shape so as to ensure a maximum influence of the agonist binding site with the ionophore channel for sodium and calcium ions. The gene for this receptor has recently been cloned and the functional regions mapped by site-directed mutagenesis and reconstitution experiments in vitro.[77] Agonist and toxin-binding sites (for α-bungarotoxin) have been identified on the α subunit and the spatial distribution of these sites in relation to the proposed three-dimensional structure of the complex determined.

The nicotinic receptor on mammalian chromaffin cells is likely to be of a similar molecular structure, though to date there has been insufficient material to characterize biochemically. In one respect it is known to differ from that at the mammalian neuromuscular junction and electroplax of the electric eel: nicotinic receptor function in chromaffin cells (release of adrenaline) is not inhibited by α-bungarotoxin.[78] However, in common with these other sources the nicotinic receptor response in isolated adrenal chromaffin cells exhibits marked tachyphylaxis (otherwise known as desensitization) in response to high concentrations of nicotinic agonists.[79]

Recent findings of interest are the apparent modulatory roles of endogenous neuropeptides on the nicotinic response, though to date nothing is known of the molecular nature of these interactions or the relation of the peptide binding sites to the agonist receptor subunit or the ionophore.

Studies by Oka et al.[80] have clarified the role of intracellular calcium in activation of the nicotinic response. These investigators found that stimulation of the nicotinic ACh responses causes an uptake of calcium that is coupled to catecholamine release and synthesis, while stimulation of the muscarinic ACh receptors causes: (1) an *efflux* of Ca^{2+} from the chromaffin cells and (2) an increase in cGMP level and incorporation of ^{32}P into phospholipids, neither of which is directly coupled to catecholamine release and resynthesis. Studies with electrically permeabilized "leaky cells"[28,81] and detergent-permeabilized cells[82-84] indicate that intracellular calcium concentrations in the range 2 to 20 μM are sufficient to initiate the process of stimulation-secretion coupling and release of catecholamines by exocytosis. The electrically permeabilized cells have also been used to study the unusual temperature sensitivity of the nicotinic response (maximally responsive to agonist at 20 to 22°C) and, as a result, to suggest that the lower response to agonist observed at higher temperatures (e.g., 37°C) is due the higher rate of nicotinic desensitization at physiological temperatures.[85] This conclusion is also supported by studies in intact isolated adrenal chromaffin cells on the temperature sensitivity of desensitization.[79]

2. What Is the Mechanism of Agonist-Induced Desensitization?

Tachyphylaxis in response to prolonged stimulation or exposure to depolarizing agents is seen not only for the release of catecholamines, but also for the release of other vesicle components such as the enkephalins. In experiments where cat adrenal glands were stimulated over a 10-min period, a rapid fatigue of output of enkephalin-like material developed during the stimulation period. During the last 2.5 min of the 10-min collection period, the release of catecholamines and enkephalins was only 25 to 33% of that in the first 2.5-min period.[68,86] The decreased output was not due to a depletion of secretory products in the gland, because the release evoked by 50 mM KCl was of equal magnitude whether or not it was preceded by perfusion of 0.1 mM ACh. This indicates that in the cat *in situ,* at these concentrations of agonists, ACh does not desensitize the secretory response to subsequent exposure to KCl. However, this is not the case in vitro with isolated bovine adrenal chromaffin cells preincubated with nicotine and then exposed to K^+.[87]

We have recently investigated this tachyphylaxis with the aim of understanding how the endocrine response of the adrenal is maintained in vivo.[87] Primary cultures of bovine adrenal chromaffin cells were exposed to successive incubation periods at room temperature and the release of norepinephrine (NE) and epinephrine (E) measured by HPLC with electrochemical detection. A 6-min "preincubation" with a desensitizing concentration of the agonist was followed by two 10-sec washes, and then a 6-min "test incubation" with a moderate concentration of the agonist. Previous exposure to a desensitizing concentration of nicotine decreased the test response to both nicotine and KCl. For example, we found that cells preincubated with 5 μM nicotine secreted 37% less NE and 24% less E in response to 50 mM KCl in the test incubation than did cells not preincubated with nicotine. Similarly, previous exposure to 50 mM KCl decreased the subsequent response to both nicotine and KCl. From this and similar experiments we concluded[87] that the loss in secretory response could not be due solely to an effect on nicotinic receptors. Nor is the loss in secretory response attributable solely to depletion of a readily releasable pool of cellular catecholamines. Cells exposed to 50 mM KCl or 5 μM nicotine in the preincubation were depleted of their cellular E by 3.4 and 6.6%, respectively, and yet the E secretion upon exposure to nicotine was reduced more by preincubation with KCl (46%) than by preincubation with 5 μM nicotine (36%).

Similar conclusions were arrived at by Boksa and Livett,[79] who observed that the cells still exhibited a loss in secretory response when exposed to agonists subsequent to a preincubation with desensitizing concentrations of agonists in a Ca^{2+}-free medium that prevented depletion of the releasable stores. Two components of desensitization were detected: (1) a

Ca^{2+}-dependent component and (2) a Ca^{2+}-independent, depletion-independent component.

Desensitization of the nicotinic response at the frog neuromuscular junction is also produced by both Ca^{2+}-dependent and Ca^{2+}-independent mechanisms.[88,89] Further support for there being two components responsible for the decline in secretion comes from studies on ox and pig adrenal glands perfused in vitro[92] where the decline persisted for a relatively long time after removal of the desensitizing agonist. If it had been exclusively a receptor desensitization process, then the perfusion without ACh between the stimulations should have allowed time for the removal of ACh from the receptors and resensitization of the response as is seen within minutes in isolated bovine adrenal chromaffin cells.[79]

The decline in secretion rate in response to prolonged exposure to agonist is, however, not simply a nicotinic receptor desensitization phenomenon as seen at the neuromuscular junction.[90] Intermittant K^{+} elevation[71] and agonists such as barium[91] and veratridine[92,93] that bypass the nicotinic receptor also elicit a declining response.

Previous work has shown that drugs that inhibit nicotinic cholinergic responses at the neuromuscular junction also affect desensitization of the nicotinic response at this synapse. For example, *d*-tubocurarine protects against nicotinic desensitization,[94] while α-bungarotoxin[95] and histrionicotoxin[95] enhance desensitization at the neuromuscular junction. *d*-Tubocurarine reduces both the rate of development and the absolute magnitude of the desensitization produced by ACh at the neuromuscular junction.

In PC12 cells, a clonal cell line derived from a rat adrenal chromaffin cell tumor, Stallcup and Patrick[96] showed that histrionicotoxin had two effects on a nicotinic response: (1) it inhibited Na^{+} uptake stimulated by the cholinergic agonist, carbachol, and (2) it enhanced carbachol-induced desensitization of Na^{+} uptake. These two effects of histrionicotoxin on receptor-mediated functions of PC12 cells were also exhibited by SP. Histrionicotoxin and SP were competitive with one another in these actions.

In cultured chromaffin cells, *d*-tubocurarine protects against desensitization of the nicotinic response (as measured by release of [^{3}H-NE] following stimulation by nicotine[8]). Interestingly, so does substance P. *d*-Tubocurarine and substance P protect against both components of desensitization (the Ca^{2+}-dependent component and the Ca^{2+}-independent, depletion-independent component), demonstrating that their ability to protect against desensitization is not a Ca^{2+}-dependent process.[8]

The realization that there is both a Ca^{2+}-dependent and a Ca^{2+}-independent, depletion-independent component of agonist-induced desensitization also provided a rationalization of published literature in this field.[79] For example, Holz et al.[10] reported that preincubation of cultured adrenal chromaffin cells for 15 min with $3 \times 10^{-4}\ M$ carbachol in the presence, but not in the absence, of Ca^{2+} inhibited subsequent carbachol-induced secretion by 86%. They interpreted these findings as indicating that carbachol-induced desensitization is Ca^{2+} dependent. However, as pointed out by Boksa and Livett[79] and Boksa,[97] the level of carbachol used ($3 \times 10^{-4}\ M$) is in the concentration range where only the calcium-dependent component of the desensitization is observed.

Likewise, the report by Schiavone and Kirpekar[98] that desensitization of the nicotinic response in perfused cat adrenal glands exposed to $5 \times 10^{-5}\ M$ nicotine is Ca^{2+} dependent is understandable in light of the more recent findings of Boksa and Livett[79] that this concentration of nicotine produced a desensitization of isolated adrenal chromaffin cells that was mainly Ca^{2+} dependent.

The exact mechanism(s) responsible for the depletion-independent Ca^{2+}-independent component of desensitization in adrenal chromaffin cells is not known, but presumably involves a step or steps distal to the binding of the agonist to the plasma membrane in the chain of events leading to stimulus secretion coupling. The most widely quoted model is that of Katz and Thesleff,[90] who proposed a slow conversion of an active to an inactive agonist-receptor complex during prolonged exposure to the agonist. More recent evidence points to a change in membrane ion conductance as a possible mechanism.[94]

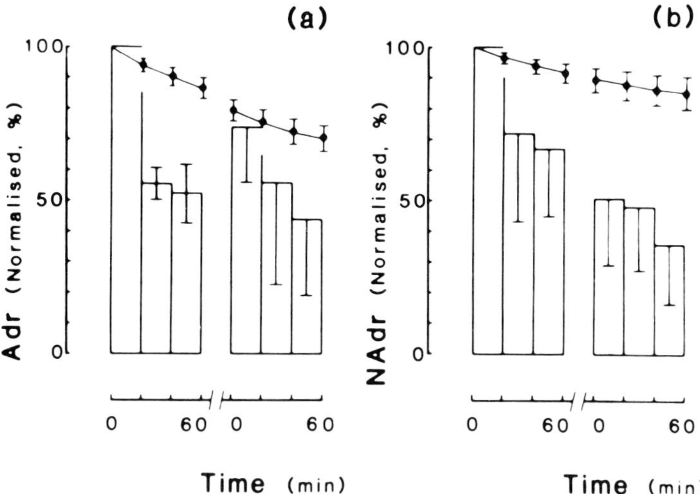

FIGURE 2.   Normalized comparison of stimulated catecholamine secretion (histogram) with catecholamine remaining in the tissue (continuous line) during infusion of 10^{-4} *M* ACh in 4 pig adrenal glands (a) for adrenaline and (b) for noradrenaline. Each gland was subjected to 1 hr of stimulation, followed by 2 hr of perfusion without ACh and then a further 1 hr of stimulation. The amount of stimulated secretion in each 20-min interval was normalized by taking the value for the first 20 min to be 100%. The mean 100% value was 330 nmol per gland for adrenaline and 480 nmol per gland for noradrenaline. The corresponding amounts of catecholamines remaining in the same four glands, at specified times, were estimated as described in text, and were normalized by taking the values at the start of the first stimulation to be 100%. The mean 100% value was 7300 nmol per gland for adrenaline and 13,400 nmol per gland for noradrenaline. Note that spontaneous (nonstimulated) catecholamine secretion has also contributed to the decline in the amount of catecholamine remaining in the tissue. (From Bevington, A. and Radda, G. K., *Biochem. Pharmacol.*, 34, 1497, 1985. With permission.)

In an attempt to examine possible links between metabolic depletion and declining catecholamine secretion, Bevington and Radda[92,99] recently carried out experiments with the perfused pig adrenal and cortex-free ox adrenal medullae stimulated by continuous infusion of 10^{-4} *M* ACh (Figure 2). Oxygen consumption and catecholamine secretion were measured in parallel to test the possibility that oxidative energy metabolism (and hence tissue receptor responsiveness) declines after prolonged exposure to agonists. However, in spite of a steady decline in the secretion of catecholamines, there was no significant decline in oxygen consumption. Hence, the decline in secretion did not result from a failure of oxidative energy metabolism.

Through these various approaches and by a process of elimination we have come to a better understanding of the phenomenon of "desensitization", even if the mechanism(s) involved remains as elusive as ever. The loss in secretory response appears to involve at least two components: (1) desensitization of the nicotinic receptor-ionophore response and, (2) inactivation of some step(s) in stimulus-secretion coupling subsequent to receptor activation, but prior to exocytotic expulsion of the vesicular contents. In addition, profound depletion of the cellular stores of catecholamines may clearly also impair subsequent secretion in response to stimuli.

B. Muscarinic Receptor Responses

The secretory response to ACh in the cat and chick[100,101] is predominantly via muscarinic receptors, whereas that in the hamster is predominantly, if not exclusively, nicotinic.[102] In other species such as bovine, dog, rabbit, rat, and man, the secretory response observed is that resulting from stimulation of both nicotinic and muscarinic cholinergic receptors. In bovine species the activation of muscarinic receptors (which occurs at ACh concentrations $<10^{-7}$ M) results in an elevation in intracellular cGMP that inhibits the nicotinic secretion of catecholamines.[103-107] On the basis of these observations it was proposed that at rest, the bovine adrenal medulla is under inhibitory (muscarinic) control, and that increased release of ACh from the splanchnic nerve terminals, as occurs normally in response to stressors such as insulin hypoglycemia, results in the nicotinic receptors being activated (ACh concentrations $>10^{-6}$) with consequent release of catecholamines.

In the adrenal medulla, as in other tissues with muscarinic receptors, the activation of these receptors is associated with an enhanced turnover of membrane phospholipids and the production of polyphosphatidylinositides.[108] What the role of these putative second messengers is in adrenal chromaffin cell secretion is not clear. In other tissues they act to mobilize intracellular calcium from the endoplasmic reticulum and, thereby, elevate cytosolic calcium levels. In bovine chromaffin cells, intracellular levels of calcium are certainly elevated prior to secretion (by nicotinic agonists) and constitute a necessary and sufficient condition for exocytotic secretion. However, muscarinic agents are ineffective in bringing about secretion in bovine chromaffin cells, although they are effective secretagogues in other species. An intriguing possibility is that the main function of the phosphatidyl inositol (PI) cycle in these cells is not to mediate the muscarinic response, but to mediate the effects of the neuropeptides upon secretion. Many neuropeptides (including substance P) that modulate the nicotinic response act through the phosphatidylinositol pathway. Further research will no doubt proceed in this direction.

C. Adrenergic Receptor Responses

Controversy seems to surround this area. On the one hand, there are reports in the literature claiming that α-adrenergic receptors[109-113] play a part in modulating secretion from adrenal medullary cells, others whose data do not exclude the possibility,[115] and still others whose data do not support such a role.[116-118] Overall, the evidence for the presence of functional α-adrenoceptors on chromaffin cells is weak.[119] More recently, Powis and Baker[120] have carried out a thorough investigation of this question using three preparations: (1) bovine adrenals perfused in a retrograde manner through the adrenal vein, (2) freshly isolated chromaffin cells, and (3) chromaffin cells maintained in tissue culture for up to 22 days. Their data come out clearly against the presence of classical α_2-adrenoceptors (either functional or silent) on chromaffin cells. Of interest, however, is that an observed inhibitory effect of the "classical" α-adrenergic agonist, clonidine, has been attributed to its ability to act as a potent antagonist of the *nicotinic* receptor.[119,120] In evidence, clonidine (IC$_{50}$ 10^{-6} M) inhibited carbachol-, ACh-, and nicotine-evoked release of catecholamines from the isolated chromaffin cells, but had no effect on the K^+ or veratridine-evoked secretion, and did not reduce the carbachol-evoked influx of $^{22}Na^+$ in the presence of ouabain and tetrodotoxin. In addition, clonidine abolished, reversibly, the ACh-induced inward current flow as detected by the patch-clamp technique.

Recent studies from our own laboratory,[235] using the ATP luminescence technique and the HPLC/ED technique for catecholamines to assess secretion, indicate that another α_2 adrenergic agonist, oxymetazoline, has a marked inhibitory effect on nicotine-induced calcium-dependent secretion, but not on K^+-induced secretion. Together, these results suggest that at least two α_2 adrenoceptor agonists act to inhibit catecholamine secretion by interaction with the nicotinic receptor-ionophore complex. This suggests that these classical α_2 adrenoceptor agonists may bind to a component of the nicotinic receptor-ionophore complex, thereby blocking Ca^{2+} movement through the receptor-linked ion channel.

Other studies have given support for the presence of β-adrenergic receptors on the chromaffin cells.[112] In bovine adrenal chromaffin cells propranolol caused a marked inhibition

(IC$_{50}$ approximately 5 × 10^{-6} *M*) of ACh-evoked catecholamine secretion,[112,118] a finding confirmed in our study using nicotine as the agonist. Propranalol did not inhibit the secretion evoked by high K$^+$. We have found that the inhibitory effect of propranalol on nicotine-evoked secretion was not competed for by isoproterenol (1 × 10^{-4} *M*) and adrenaline,[235] indicating that as with clonidine inhibition of secretion this β-adrenergic receptor blocker may exert its inhibitory effect on catecholamine secretion via the nicotinic receptor-ionophore complex.

In summary, while much is known about the relative potency of different agonists and antagonists and the different ionic events initiated, little is known in detail about how these changes in ion levels bring about the final exocytotic event. Theories discussed elsewhere in this volume implicate the cytoskeleton and actomyosin-like contractile elements, as elements sensitive to the raised low levels of intracellular calcium accompanying stimulus-secretion coupling.

III. NEUROPEPTIDE MODULATION OF SECRETION

One of the first studies to look at the possible role of endogenous peptides on adrenal medullary secretion was that of Staszewska-Barczak and Vane,[121] who investigated the role of angiotensin II, brandykinin, eledoisin, and kallidin on the *basal* secretion of catecholamines from the adrenal medulla. These peptides had little effect on the basal secretion of catecholamines and this line of investigation was not taken up again until the discovery of endogenous peptides in the adrenal in the mid 1970s.

A. Neuropeptides in the Adrenal Medulla and Splanchnic Nerve

It has been known for 40 years that the adrenal chromaffin granules contain large amounts of soluble proteins (the chromogranins)[70] packaged with the catecholamines and ATP. Helle[122] drew attention to the fact that in addition to the major chromogranins (chromogranin A, chromogranin B, and dopamine-β-hydroxylase), there were some 22 unidentified proteins and peptides of smaller molecular weight. However, it was not appreciated until quite recently that these may be the source of biologically active peptides. Further, it was not appreciated until very recently that the splanchnic nerve that innervates the adrenal contains, in addition to the cholinergic motor fibers, sensory fiber collaterals rich in SP, and other fibers of unknown origin that contain other neuropeptides (e.g., opioid peptides).[52,123] There is reason to believe that at least some of the cholinergic nerves comprising the splanchnic innervation to the adrenal medulla costore neuropeptides with acetylcholine.[52,124]

Table 1 summarizes what is known of the occurrence and structure of neuropeptides known to be present in the adrenal medulla and splanchnic nerves of various species.

1. Opioid Peptide Localization

Enkephalin-immunoreactive material has been localized in association with adrenaline (but not noradrenaline) cells in tissue sections of the bovine adrenal medulla[125] and in the cytoplasm and processes of bovine chromaffin cells in culture.[126-128] In bovine chromaffin cells in monolayer culture the opioid peptides are concentrated within the varicose processes, suggestive of transport and processing of the opioids as they move along the axons.[126] It is not known whether these enkephalin-containing cells are exclusively adrenaline synthesizing, but release studies[129] indicate a parallel release of enkephalin with adrenaline.

Subcellular fractionation studies[130] have shown that the bulk of the total Met-enkephalin immunoreactivity (42%) is recoverable in the large granule fraction that also contains 38% of the total dopamine β-hydroxylase activity and 42% of the total catecholamines. When the chromaffin granule fraction was further fractionated to give a fraction containing dense noradrenaline vesicles and another less dense fraction rich in adrenaline-containing vesicles,

Table 1
AMINO ACID SEQUENCES OF SOME NEUROPEPTIDES ENDOGENOUS TO THE ADRENAL MEDULLA OR ITS INNERVATION

Amino acid sequences of some enkephalin congeners[a]
  Pro-enkephalin A[b]

Met-enkephalin	**YGGFM**
	YGGFMRF
	YGGFMRGL
Leu-enkephalin	**YGGFL**
Metorphamide (adrenorphin)	**YGGFMRRV**-NH$_2$
BAM-12-P	**YGGFMRRVGRPE**
BAM-20-P	**YGGFMRRVGRPEWWMDYQKR**
BAM-22-P	**YGGFMRRVGRPEWWMDYQKRYG**
Peptide E	**YGGFMRRVGRPEWWMDYQKRYGGFL**
Peptide F	**YGGFMKKMDELYPLEVEEEANGGEVLGKRYGGFM**
Amidorphin	**YGGFMKKMDELYPLEVEEEANGGEVL**-NH$_2$

  Pro-enkephalin B[b]

a-Neoendorphin	**YGGFLRKYPK**
β-neoendorphin	**YGGFLRKYP**
Dynorphin 1-17	**YGGFLRRIRPKLKYDNQ**
Dynorphin 1-8	**YGGFLRRI**
Rimorphin	**YGGFLRRQFKVVT**

  Pro-opiomelanocortin
   (only in human adrenal medulla)

β-endorphin	**YGGFM**TSEKSQTPLVTLFKNAIIKNAYKKGQ

Substance P and related tachykinins[c]
  Mammalian

Substance P[b]	**RPKPQQFFGLM**-NH$_2$
Neurokinin A	HKTDS**FVGLM**-NH$_2$
Neurokinin B	DMHD**FFVGLM**-NH$_2$

  Amphibian

Kassinin	DV**PK**SDQ**FVGLM-NH$_2$**
Physalaemin	pEAD**PNKFYGLM-NH$_2$**
Uperolein	pEPD**PNAFYGLM-NH$_2$**
Phyllomedusin	pEN**PNRFIGLM-NH$_2$**

  Molluscan

Eledoisin	pEPSKDAFIGLM-NH$_2$

Amino acid sequences of VIP- and NPY-releated peptides[c]
  VIP-related peptides

pVIP[b]	HSDAVFTDNYTRLRKQMAVKKTLNSILN-NH$_2$
pPHI	HADGVFT**SDFS**RLLGQLSAKKTLESLI-NH$_2$
hPHM	HADGVFT**SDFS**RLLGQLSAKKTLESLM-NH$_2$

  NPY-related peptides

pNPY[b]	YPSKPDNPGEDAPAEDLARYYSALRHYINLITRQRY-NH$_2$
pPYY	YPAKPEAPGEDA**SPEELS**RYYA**SL**RHYLNLVTRQRY-NH$_2$
APP	**GPSQPTY**PGDD**APVEDLIRFYDNLQQYLNVV**TRHRY-NH$_2$
BPP	**APLEPEY**PGDN**ATPEQMAQYAAEL**RRYINMLTRPRY-NH$_2$
HPP	**APLEPVY**PGDN**ATPEQMAQYAADL**RRYINMLTRPRY-NH$_2$

Some other endogenous peptides with receptors in the adrenal medulla[d]

bSomatostatin[b]	AGCKNFFWKTFTSC
fBombesin	pEQRLGNQWAVGHLM-NH$_2$
bNeurotensin[b]	pELYENKPRRPYIL
hAngiotensin II	DRVYIHPF

Note: The following is the one-letter amino acid code: A = Ala, C = Cys, D = Asp, E = Glu, F = Phe, G = Gly, H = His, I = Ile, K = Lys, L = Leu, M = Met, N = Asn, P = Pro, Q = Gln, R = Arg, S = Ser, T = Thr, V = Val, W = Trp, Y = Tyr.

Table 1 (continued)
**AMINO ACID SEQUENCES OF SOME NEUROPEPTIDES ENDOGENOUS TO
THE ADRENAL MEDULLA OR ITS INNERVATION**

[a]   Bold type indicates the common pentapeptide enkephalin sequences.
[b]   Peptides endogenous to the adrenal medulla. For further details, see Marley and Livett.[6]
[c]   Species: p = porcine, h = human, APP, BPP, and HPP = avian, bovine, and human pancreatic polypeptides, respectively. Bold type indicates residues differing from VIP or NPY.
[d]   Species: b = bovine, f = frog, h = human.
[e]   Bold type indicates common tachykinin sequences.

the total Met-enkephalin immunoreactivity was restricted to the fraction containing the lighter adrenaline-containing vesicles. These findings are supported by the findings of Lang and colleagues[131] who found Leu-enkephalin immunoreactivity to be absent from the dense noradrenaline vesicles.

Exclusive localization of the enkephalins to the adrenaline-containing cells may not be a general species phenomenon, since quite different conclusions were arrived at in another study[132a] involving ultrastructural localization of Met[5]enkephalin using a gold-labeled antibody technique in the adrenal medulla of the cat, pig, and man. These investigators found no evidence for a selective association of the enkephalins with adrenaline or noradrenaline cell types. Rather, they found that the majority of granules of each cell type exhibited Met[5]enkephalin-like immunoreactivity. The enkephalins were also colocalized with dopamine β-hydroxylase in electron-dense secretory granules in human pheochromocytomas, and in the carotid bodies from the cat, pig, and man. Unfortunately, these investigators did not report the result of any findings with this technique for bovine species that might have served as a useful comparison. Another study[132b] found Met-enkephalin-Arg[6]Phe[7] and Met-enkephalin-Arg[6]Gly[7]Leu[8]-like immunoreactivity in dog, cat, and rat adrenal adrenaline and noradrenaline cells with no selective localization in these three species, whereas in bovine species *in situ* hybridization[132c] has shown selective expression of proenkephalin-A mRNA in the adrenaline cells, thus confirming the immunocytochemical data.[125]

In the adrenal medulla of all species so far studied, the dynorphins[105,133-135] are present in much lower amounts than the processed enkephalins and appear to be associated with the noradrenaline-containing cells.[34,136-139a] A number of different molecular forms of dynorphin have been identified by fast-atom bombardment/mass spectrometry of peptide fragments obtained from bovine adrenal medulla. These include Dyn (1-11), Dyn (1-12), Dyn (1-13), and two modified forms of Dyn (1-13) with molecular weights exceeding that of Dyn (1-13) by 14 and 28 amino acids, respectively. Upon stimulation of bovine adrenal chromaffin cells with acetylcholine, immunoreactive (ir) dynorphins are released into the medium together with the catecholamines and enkephalins. There is evidence for a preferential secretion of enkephalins from cells enriched in adrenaline and of ir-Dyn from adrenal chromaffin cells enriched in noradrenaline.[139a]

The functional significance, if any, of this preferential localization of the processed enkephalins and dynorphins with adrenaline- or noradrenaline-containing cells in the bovine and hamster[139b] adrenal, but not apparently in the cat, pig, or man, is not known, but may be due to the use of antisera with different specificities. It is possible that both proenkephalins are expressed in both cell types. The precursors may be processed down to the pentapeptide enkephalins in adrenaline cells, but only to larger products, such as BAM-22P and dynorphins, in noradrenaline cells. The presence of the opioid peptides in one or another of the two cell types would then depend on the staining specificity of the antisera used for immunocytochemistry. On the other hand, the finding of selective expression of proenkephalin-A mRNA in adrenaline cells of the bovine adrenal[132c] suggests that in this species there may be real differences in enkephalin expression in adrenaline and noradrenaline cell types.

2. SP Localization

SP has been detected by radioimmunoassay in adrenal medullary tissue,[6,74,140-147a] and in small amounts in adrenal chromaffin cells isolated from the adrenal medulla of the cow (38 to 919 fmol/10^6 cells[236]) and guinea pig (0.8 pmol/mg protein).[146,147a] Most of the immunoreactive SP appears to be located in nerve terminals of the splanchnic nerve innervating the medulla, as seen in immunohistochemical studies on human[123] and rat[148] adrenal glands. In the rat adrenal an occasional SP-containing cell has been seen in the areas between islets of chromaffin cells.[148]

Little is known of the origin of the SP-containing fibers within the splanchnic nerve. It is possible that they represent collaterals of primary afferent neurones, similar to those that innervate the inferior mesenteric ganglion and that have their cell bodies in spinal ganglia.[149-151] Support for this idea comes from the work of Gamse et al.[140] who showed that the SP-like immunoreactivity in the adult guinea pig splanchnic nerve was depleted by 83% following neonatal capsaicin pretreatment. This finding suggests that SP is located in sensory fibers. A similar depletion of SP-like immunoreactivity was obtained in the rat splanchnic nerve following neonatal capsaicin treatment of Sprague-Dawley[152] (80% depletion) and Buffalo (70% depletion)[153] strains. Alternatively, the SP-containing fibers in the adrenal may be terminals of preganglionic neurones or of cell bodies located in the gut producing efferent projections. However, there is presently no evidence for such SP-containing neurones.[6]

These findings raise the possibility that control of adrenal medullary secretion in vivo may involve these endogenous neuropeptides.

B. Role of Neuropeptides as Modulators of Nicotinic Function: The Concept of Neuromodulation

Although there is presently no recognized definition of neuromodulation, two universally accepted hallmarks of a neuromodulator are (1) it should show contingent action and (2) its action(s) should be relatively long lasting. These two properties of a neuromodulator are shown best by SP in its actions on a number of nicotinic receptor systems, as reviewed below. The review that follows will concentrate mainly on the nicotinic receptor actions of tachykinins and opioid peptides on the adrenal, since these two classes of neuropeptides have been the most studied. However, some studies have also been made with somatostatin, neurotensin, VIP, and other endogenous peptides, and, where appropriate, these will also be considered.

1. Effects of Endogenous Opioid Peptides on Secretion

Although the adrenal medulla represents one of the richest stores of opioid peptides in the body, there is still no clear evidence that the opioid peptides endogenous to the adrenal medulla play a role in controlling secretion from the chromaffin cells in vivo. Studies with isolated adrenal chromaffin cells in vitro have shown that the two principal opioid pentapeptides, Leu^5- and Met^5-enkephalin, are poor modulators of the nicotinic response (EC_{50} > 10^{-4} M).[154,155] The potency increases with increasing chain length so that β-endorphin is one of the most potent opioid peptides in vitro (EC_{50} 10^{-6} M). However, β-endorphin is not present in the adrenal medulla of any mammalian species studied, except man, and its levels within the adrenal and plasma are extremely low, so it is hard to imagine how it might be effective in vivo. Of some interest, however, is the recent finding that several larger MW congeners of the enkephalins are normally present as processed cleavage products of the endogenous precursors, proenkephalin A and B (prodynorphin). As reviewed in detail later, two of these, metorphamide[156] (otherwise known as adrenorphin, eight amino acids; from proenkephalin A) and dynorphin (1-13) (from proenkephalin B), are potent inhibitors of the nicotinic response in isolated adrenal chromaffin cells (IC_{50}s 5 × 10^{-6} M, 3 × 10^{-6}

M, and 6 × 10⁻⁶ *M,* respectively[56,154]). Moreover, a nonopioid portion of these peptides is active at inhibiting the nicotinic response, suggesting multiple biological activities may be present within the one molecule, as is well known for proopiomelanocortin.

The opioid peptides discussed above have been reported to have a number of effects upon secretion: (1) at high concentrations, they enhance the basal secretion of catecholamines; (2) at a range of concentrations, depending on the particular peptide concerned, they inhibit nicotinic-mediated secretion; and (3) some of them have now been tested for their effects on desensitization of the nicotinic receptor and, like SP, they have been shown to have a protective role against nicotinic desensitization. The presence of these opioid peptides in relatively high concentrations in the chromaffin cells and in the splanchnic innervation of the adrenal medulla suggests a functional role for them in vivo.

In all, over 30 different molecular forms of the opioid peptides have been found in the mammalian adrenal medulla, either in chromaffin cells, their innervation, or both.[6,157,158] There is evidence that the complete processing of proenkephalin in the adrenal medulla involves a number of sequential proteolytic cleavages starting at the C-terminal end of the molecule.[156-159] Commonly, the active opioid is flanked by pairs of basic amino acids in the precursor so that sequential action of a trypsin-like endopeptidase followed by a exopeptidase would liberate the hormone. Recent studies on the carboxypeptidase-B-like processing enzyme associated with enkephalin biosynthesis in the chromaffin granules[160] indicate that this enzyme may also be involved in the processing of a number of other peptide hormones and neurotransmitters. Similarly, an amidating enzyme isolated from bovine anterior and intermediate lobe secretory granules converts many peptides with a C-terminal glycine into the corresponding amide.

Whether unique enzymes are associated with each prohormone, or whether general processing enzymes are involved in producing similar cleavages in many prohormones is not known, however, the processing of proenkephalin does appear to be tissue specific.[161-163] Two novel high molecular weight enkephalin-containing peptides (thought to be intermediates in the processing of proenkephalin A) have been isolated recently from the bovine adrenal medulla using immunoblotting combined with specific RIAs.[164] One contains the Met-enkephalin-octapeptide sequence at its C terminus and appears larger than 18.2 kdaltons. The other extends further toward the C-terminal end of proenkephalin, terminating with the sequence of Leu-enkephalin. The processing enzymes responsible for these specific cleavages have not been identified with certainty, although it is known that several of the larger peptides are substrates for tissue kallikreins.[165]

These large molecular weight enkephalin-containing peptides are coreleased with Leu- and Met-enkephalin from the adrenal and other tissues.[86,166-168] In contrast to Met-enkephalin, which is rapidly destroyed when released, synenkephalin, the 70 amino acid, cysteine-rich, N-terminal part of bovine proenkephalin A is not destroyed.[167] Several of the larger enkephalin-containing peptides isolated from the bovine adrenal medulla contain synenkephalin at their N terminal.[161,169,170]

In cat adrenals only 25% of the enkephalin-like material is there as free Met⁵enkephalin:[86] most of it is in the form of large enkephalin-containing peptides. Upon electrical stimulation (15 Hz) of the splanchnic nerve *in situ,* Met-enkephalin and its hepta- and octapeptide congeners are released in molar ratios of 4:1:1, which is the same ratio as they are found in the precursor, proenkephalin A. However, further analysis revealed that the nature of the released peptides depends on the type of stimulus used to evoke release. While electrical stimulation (15 Hz) of the splanchnic nerve, or perfusion of the adrenal with Krebs solution containing ACh (0.1 m*M*), or KCl (50 m*M*) for 10 min, all induced an immediate release of the three opioid peptides that mirrored the output of catecholamines, perfusion with ACh or KCl resulted in a proportionately greater amount of enkephalin-containing peptides of larger molecular weight. This difference in composition of the released materials has been taken to indicate that under physiological conditions most of the enkephalin-immunoreactive materials that are released are the final products obtained as a result of full processing of proenkephalin, namely, Met[5]-enkephalin, Leu[5]-enkephalin, Met[5]-enkephalin-Arg[6]-Phe[7], and Met[5]-enkephalin-Arg[6]-Gly[7]-Leu[8], whereas when high concentrations of depolarizing agents are used, partially processed materials are also released.[86,166] It is thought likely that physiological stimulation causes preferential release of "mature" granules containing fully processed enkephalins, whereas high concentrations of depolarizing agents probably mobilize, in addition, some "immature" granules containing partially processed materials.

The possible actions of the larger opioids are not known, however, several of the enkephalin congeners have a higher potency than Leu[5]- or Met[5]-enkephalin themselves. Both the heptapeptide and the octapeptide interact with high affinity with the opioid receptor in in vitro binding studies, and together with the BAM peptides (BAM-12P, BAM-20P, BAM-22P) are potent agonists in the mouse vas deferens and guinea pig ileum,[55,171,172] suggesting that they may function normally as neurotransmitters or neuromodulators.

Recently, a new opioid peptide was characterized from both a human pheochromocytoma (and termed "adrenorphin")[173] and from bovine striatum ("metorphamide").[174] Comparison of the amino acid sequences of the two peptides shows them to be identical. Metorphamide (adrenorphin) consists of the sequence of Met-enkephalin with an amidated three amino acid extension at its C terminal (Met-enkephalin-Arg-Arg-Val-NH$_2$). Metorphamide was also reported to be present in high concentrations in normal adrenal tissue.[174] This octapeptide sequence is contained within the sequence of three of the larger adrenal opioid peptides, BAM-12P, BAM-22P, and peptide E (of 12, 22, and 25 residues, respectively), which are in turn contained within the sequence of proenkephalin A. Although reported to be present in high concentration in the adrenal medulla, little is known of the distribution of metorphamide in the adrenal medulla and splanchnic nerve. Being a proenkephalin A-derived product, it is, like Met-enkephalin,[125,175] probably contained preferentially within the adrenaline cells of the adrenal medulla.

Marley et al.[56,176] recently tested the ability of metorphamide to modify the nicotine-induced exocytosis from isolated bovine adrenal chromaffin cells in vitro in two different systems: fresh, dispersed adrenal medullary cells, and primary monolayer cultures of these cells maintained in culture for 3 days. Secretion was induced with 5 μM nicotine and was monitored by measuring ATP secretion from the fresh cells (by the luciferin-luciferase bioluminescence method[177]) or by measuring endogenous catecholamine secretion from the cultured cells (using HPLC with electrochemical detection[177]). Metorphamide produced a dose-dependent inhibition of catecholamine secretion in the range 1 to 20 μM (IC$_{50}$ = 11 μM), but had no effect in lower concentrations (1 nM to 1 μM). The inhibition of adrenaline and of noradrenaline secretion was identical and amounted to 85% at 20 μM, the highest concentration tested. Metorphamide was more effective at inhibiting secretion from fresh cells (IC$_{50}$ = 5 μM) than from cultured cells, and its activity was not increased by bacitracin, 1 mM; bestatin, 15 μM; thiorphan, 0.1 μM; or 1,10-phenanthrolene, 0.1 mM. Of note, metorphamide was some 44-fold more potent than Leu-enkephalin, and 100-fold more potent than Met-enkephalin[56] at inhibiting catecholamine secretion from the isolated chromaffin cells. Moreover, it was the only opioid peptide containing the sequence of Met-enkephalin of ten tested (including BAM-12P and BAM-22P which contain the metorphamide sequence, but lack the C-terminal amide[54]) that was active at 5 μM (Figure 3).[155]

It is proposed that the novel amidated C-terminal valine of metorphamide is important in conferring potency toward the nicotinic receptor, since the free acid ("metorphamate"[237]) or removal of the C-terminal Val-amide residue (to generate Met[5]-enkephalin-Arg[6]-Arg[7]), or the hexa- and pentapeptides are much less potent than metorphamide.[56,154,176] However, degradation of the peptides is unlikely to be the main cause for their apparent lack of potency, for, whereas the potency of dynorphin (1-9) and dynorphin (1-13) was enhanced by about

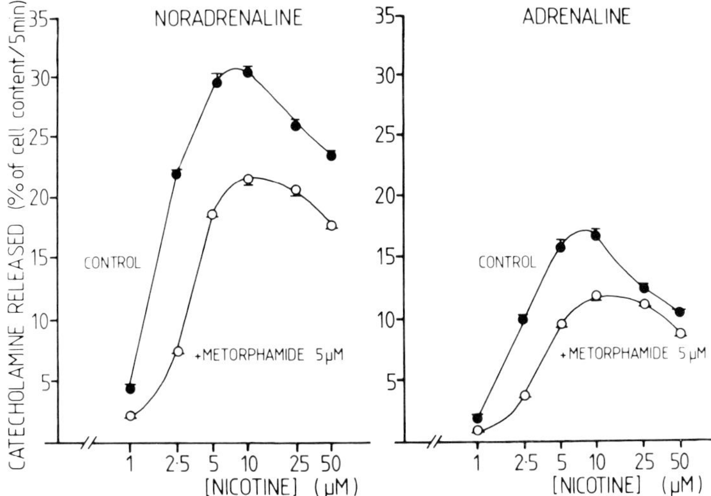

FIGURE 3.   Inhibition by 5 μ*M* metorphamide of endogenous noradrenaline (left panel) and adrenaline (right panel) release induced by different concentrations of nicotine. Increasing concentrations of nicotine were not able to overcome the inhibitory effect of metorphamide, suggesting that metorphamide acts noncompetitively with respect to the actions of nicotine at the nicotinic receptor. (From Marley, P., Mitchelhill, K. I., and Livett, B. G., *Brain Res.*, 363, 10, 1986. With permission.)

20% by the protease inhibitors bestatin or thiorphan, that of rimorphan and of seven other Met[5]-enkephalin-containing opioid peptides tested, including metorphamide, was not.[155] As with the other less potent opioid peptides, the inhibitory actions of metorphamide were unaffected by naloxone or diprenorphine, suggesting that they were mediated through a class of receptors distinct from those that mediate high-affinity opiate binding as characterized by others using ligand-binding studies.

An even larger Met[5]-enkephalin-containing opioid peptide possessing a C-terminal amide has been isolated recently from the bovine adrenal medulla. This peptide of 20 amino acids, termed amidorphin, is also reported to be present in the bovine gland in considerable quantities, but to date has not been tested for its activity toward the nicotinic receptor. Even if it is active against nicotinic receptor function in isolated bovine chromaffin cells, there is no certainty that this will turn out to be a physiologically important mechanism, since to date the bovine species is the only one to exhibit this second endogenous amidated peptide (the -Gly-Lys-Arg- bond so essential for forming the C-terminal amide on amidorphin from bovine proenkephalin A is not contained within the structure of prokephalin A from rat, human, or frog).

Studies from a number of laboratories support the view that chromaffin cells possess several distinct opioid ligand binding sites that display stereoselectivity between isomers.[154,178-184] One of these opioid recognition sites has the unusual property of recognizing etorphine, β-endorphin, and Met[5]enkephalin-Arg[6]-Phe[7] with high affinity and morphine, naloxone, and Met- and Leu-enkephalin poorly.[180,182-185] Another has been found to be an unusual κ-site with a high affinity for Met[5]-enkephalin-containing peptides.[186]

Given the unusual characteristics of the opioid binding site(s) in the adrenal medulla[180,181,186] and the inability of naloxone to prevent the inhibitory actions of metorphamide,[56] Met-enkephalin, and dynorphin (1-13)[154] on the nicotinic response, it is likely that these peptides produce their effects either by interacting with naloxone-resistant opioid receptors or through receptors that recognize the nonopioid parts of the peptides. The latter possibility is supported by our recent observation that N-terminal fragments of dynorphin (1-13)[155] and metorphamide[237] are much less potent or inactive in this preparation, while C-terminal fragments retain activity. Whatever the mechanism it is clear that opioid receptors of the classical kind are not involved in mediating the inhibitory modulation of the nicotinic response. This view is supported by the poor opiate stereoselectivity between dextorphan and levorphanol in inhibiting the nicotinic response[154,184] and by the observation that opioid peptide analogs such as diiodo-Tyr[1]-β-endorphin and diiodo-Tyr[1]-dynorphin (1-13), both of which are inactive in bioassays and in opioid-binding assays in classical opioid receptive tissues, are equipotent with their respective endogenous opioids in inhibiting the nicotinic response in chromaffin cells.[154]

2. Effects of Tachykinins on Secretion

There were precedents for believing that SP may have a role in the control of nicotinic function in the adrenal medulla. An inhibitory modulatory role for the peptide SP had been reported for the nicotinic receptor on Renshaw cells in the cat spinal cord.[187-191] In these electrophysiological studies, SP was found to decrease the acetylcholine-evoked nicotinic activation of Renshaw cells without affecting the activation caused by aspartate, glutamate, or muscarinic agonists. Subsequent work on the Mauthner fiber/giant fiber synapse in the hatchetfish,[192] the peripheral neuromuscular junctions in the bullfrog,[193,194] and the carotid chemoreceptor in the cat[195] has extended the generality of these inhibitory effects of SP to the nicotinic receptor in nonvertebrates and to other systems. In its actions on the adrenal medulla and in a number of neuronal systems, SP does not act as a typical neurotransmitter causing depolarization of the postsynaptic membrane and generation of action potentials, but rather by modifying the responsiveness of a postsynaptic membrane receptor to another transmitter. This "modulatory" action of SP is well recognized[196] and the concept of neuromodulation has been invoked in a number of other neurotransmitter systems.[197]

a. Substance P and Analogs

SP has two distinct actions on the nicotinic response in cultured bovine adrenal chromaffin cells: namely, inhibition of agonist-induced secretion and protection against nicotinic desensitization.

i. SP Inhibition of the Nicotinic Response

SP inhibits nicotine- and ACh-evoked secretion of catecholamines, but has no effect on basal release or on the release induced by 56 mM K$^+$, veratridine, or sucrose substitution of NaCl.[8,147,198] SP also has no effect on the release induced by Ca^{2+}-ATP from "leaky" chromaffin cells.[199]

In these actions SP appears to interact with a regulatory site on the nicotinic receptor-ionophore complex. Studies by Boksa et al.[200] with SP analogs having single amino acid substitutions have shown that SP receptors on the cultured chromaffin cells exhibit similar structural requirements for activation, as do SP receptors in other SP-responsive tissues.

ii. Protection Against Nicotinic Desensitization

The second action of SP in the adrenal is that it protects against ACh- and nicotine-induced desensitization of the nicotinic response.[8,54] SP ($>5 \times 10^{-6}$ M) completely protects against desensitization produced by ACh ($>10^{-4}$ M) or nicotine ($>2.5 \times 10^{-6}$ M), but has no effect on K$^+$-induced desensitization (Figure 4). SP protects against both depletion of catecholamine stores and a depletion-independent component of desensitization.[79] This protection against desensitization is a result of facilitation of catecholamine release rather than inhibition of catecholamine reuptake.[8,54]

iii. Studies with Pheochromocytoma Cells

Similar studies to those described above have been carried out on PC12 cells (a cell line derived from a rat adrenal noradrenaline chromaffin cell tumor), to show that: (1) SP inhibited

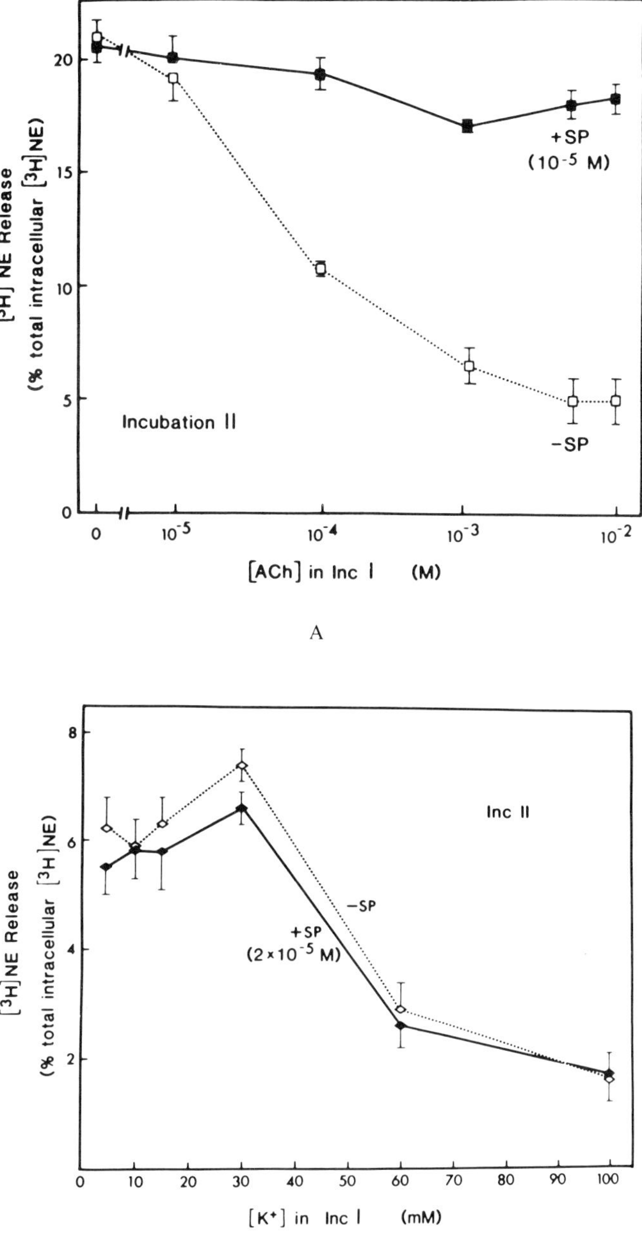

FIGURE 4. (A) Protection against desensitization of the nicotinic response by SP. The effects of SP on ACh-induced desensitization of [³H]NE release at 37°C are shown. Chromaffin cell cultures were preincubated at 37°C for 5 min with the range of ACh concentrations shown, ±SP (10^{-5} *M*), according to the protocol described in Figure 1 of Boksa and Livett,[8] and then washed twice with KRH buffer. The cultures were then restimulated with an EC_{50} concentration of ACh, and the [³H]NE released during this second incubation (Inc II) is shown. Each point represents the mean (±SEM) of the results from three cultures. (B) Lack of effect of SP on K⁺-induced desensitization of [³H]NE release. Chromaffin cell cultures at 23°C were preincubated for 5 min with the range of K⁺ concentrations indicated, in the presence and absence of 2×10^{-5} *M* SP (Inc I, the preincubation). Cultures were then rapidly washed and restimulated for 5 min with 60 m*M* K⁺ (Inc II). The [³H]NE released during Inc II is expressed as a percentage of the total intracellular [³H]NE present in each culture at the beginning of the incubation period. Each point represents the mean (±SEM) of results from three cultures. (From Boksa, P. and Livett, B. G., *J. Neurochem.*, 42, 618, 1984a. With permission.)

[22]Na[+] uptake stimulated by the cholinergic agonist, carbachol; and (2) SP *enhanced* carbachol-induced desensitization of [22]Na[+] uptake and stabilized the desensitized state of the receptor.[96] At first sight, these results appear in contrast with those obtained with bovine adrenal chromaffin cells in culture[8] in which SP *protected* against nicotinic receptor desensitization. However, these two apparently contradictory effects of SP on desensitization might be reconciled if Na[+] uptake was not linked to catecholamine release or to desensitization of catecholamine release, i.e., SP could then be seen as having opposing effects on two independent nicotinic responses. This appears to be the case, since extracellular Na[+] is not an obligatory requirement for cholinergic release of catecholamines from adrenal chromaffin cells,[201] and catecholamine release can be effectively desensitized by nicotine in the absence of extracellular Na[+].[8]

The ability of SP to inhibit noradrenaline release is, however, influenced by Na[+] as evidenced by the observation that the release was greater in standard (125 mM Na[+]) than in reduced Na[+] medium (i.e., desensitization can occur in conditions where agonist-induced release does not occur). Although the mechanism of this action is not known, it is conceivable that Na[+] might decrease the affinity of SP for its receptor as has been demonstrated for other peptide receptors.[202] Similarly, in the electrophysiological study of bovine adrenal chromaffin cells using the patch-clamp technique[203] and in a related study in cultured chick sympathetic and ciliary ganglion neurons by the voltage-clamp technique,[204] the action of SP to increase the rate of decay of ACh or carbachol-induced ion current occurred over a time frame of 1 to 5 sec. In contrast, the time frame over which catecholamine release takes place from the isolated chromaffin cells in vitro is 5 min. Hence, the initial events in nicotinic receptor activation by ACh may involve an enhanced rate of receptor-linked Na[+] ion movement by SP, but it is by no means certain that this early response is related to the subsequent secretory response of catecholamines that is maintained following preincubation with substance P.

iv. Studies with Other Tachykinins and the Nature of the SP Receptor Subtype in the Adrenal

As to the molecular mechanism(s) of SP action at the nicotinic receptor-ionophore complex, little is known other than that SP appears to interact noncompetitively to regulate Na[+] entry. SP appears to interact with a site on the chromaffin cells distinct from those recognized by nicotine, or by local anesthetics such as procaine, amantidine, and QX222 that compete for Na[+] entry. Rather, it appears to interact noncompetitively with a distinct modulatory site similar to that recognized by quinacrine and histrionicotoxin.[53,96,205] In view of the known actions of cholinergic muscarinic agents to elevate cGMP levels in bovine chromaffin cells and inhibit the nicotinic response,[103] it would be of interest to see if SP brings about its inhibitory effects by elevating cGMP, as has been shown for the action of SP in isolated pancreatic acinar cells.

Calcium ions are unlikely to be second messengers for the actions of SP at the nicotinic receptor, for, unlike secretion from parotid[206] and salivary glands[207] where SP action is Ca^{2+} dependent, its protection against nicotinic receptor desensitization in chromaffin cells is independent of external Ca^{2+}.[8,97] Recent studies by Boksa[97] have examined this question in detail. She found that SP modulates two other cholinergic responses, [45]Ca^{2+} uptake and [45]Ca^{2+} efflux. Once again, two effects of SP were observed: (1) SP inhibited carbachol-induced [45]Ca^{2+} uptake and [45]Ca^{2+} efflux, and (2) SP protected against desensitization of carbachol-induced [45]Ca^{2+} uptake and [45]Ca^{2+} efflux. Of relevance to the studies reviewed above, the ability of SP to inhibit carbachol-induced [45]Ca^{2+} uptake was noncompetitive with respect to Ca^{2+} and increasing concentrations of extracellular Ca^{2+} were unable to overcome SP inhibition of [3H]-NE release. The results also indicated that SP inhibition of net carbachol-induced [45]Ca^{2+} uptake was due to inhibition of [45]Ca^{2+} uptake rather than enhancement of [45]Ca^{2+} efflux. SP blocked the carbachol-induced uptake of [45]Ca^{2+} in both Na[+]-containing and Na$^+$-free media, however, SP produced only a small inhibition of K$^+$-induced $^{45}Ca^{2+}$ uptake, indicating that SP does not interact directly with voltage-sensitive Ca^{2+} channels.

Although endogenous intracellular Ca^{2+} levels were not measured in this study, Boksa[97] argues that it is unlikely that the ability of SP to inhibit carbachol-induced $^{45}Ca^{2+}$ uptake was due to inhibition of $^{45}Ca^{2+}$-Ca^{2+} exchange because (1) the rate of $^{45}Ca^{2+}$-Ca^{2+} exchange in the adrenal medulla is known to be dependent on the extracellular Ca^{2+} concentration,[208] yet alterations in extracellular Ca^{2+} had almost no effect on the SP inhibition of $^{45}Ca^{2+}$ uptake induced by carbachol, and (2) the release of catecholamines is Ca^{2+} dependent and triggered by a rise in intracellular free Ca^{2+}. Rather, the similarities in the effects of SP on carbachol-, nicotine-, and ACh-induced release of [^{3}H]-NE and on $^{45}Ca^{2+}$ uptake suggest that SP action may be directed primarily at modulating endogenous *intracellular* calcium levels. This hypothesis should be directly testable now that a range of intracellular Ca^{2+}-sensitive dyes are available for experimentation.[28]

Due to considerable interest in SP actions over the last 15 years resulting from the early publication of its structure and chemical synthesis by Tregear and Leeman in 1967,[209] much is known about structure-activity profiles of SP analogs, congeners, and related tachykinins in various tissues. As a result, three major classes of receptor interactions have been described:[210] (1) those typified by physalaelmin acting on the guinea pig ileum ("SP-P" receptors), (2) those typified by eledoisin acting on the rat vas deferens ("SP-E" receptors), and (3) those typified by neuromedin K acting on rat brain cortical membrane preparations ("SP-K" receptors). These three tachykinin receptor subtypes "SP-P", "SP-E", and "SP-K" are proposed to have as endogenous ligands SP, neuromedin K (NK), and substance K (SK) (see Table 1).

Structure-function studies on the SP receptor subtype in bovine adrenal chromaffin cells have shown that the rank order of potencies of physalaemin, kassinin, eledoisin, SP, and SP methyl ester are different from those reported for the guinea pig ileum or rat vas deferens.[211] The naturally occurring tachykinins, physalaemin, eledoisin, and kassinin, were about equipotent and much less potent than SP in inhibiting nicotine-induced ^{3}H-NE release. The rank order of potency of these and other SP-related peptides for inhibition of ^{3}H-NE release was similar to their rank order for protection against desensitization of the nicotinic response. The chromaffin cell SP receptors mediating inhibition of nicotine-induced ^{3}H-NE release appear to be very similar to those mediating protection against desensitization. These SP receptors are readily activated by SP-methyl ester and SP, but are only poorly activated by SP-free acid and the naturally occurring tachykinins, physalaemin, eledoisin, and kassinin. D-Pro^2,D-Trp^{7,9}-SP did not antagonize chromaffin cell SP receptors. Although it is likely that SP modulates adrenal catecholamine release in vivo, these findings suggest that a kassinin- or physalaemin-like peptide probably does not.

More recent studies utilizing HPLC/ED to detect endogenous catecholamine release, and the firefly luminescence technique to detect ATP release from cultured bovine adrenal chromaffin cells have confirmed these results with ^{3}H-NE release and, in addition, have shown that neither neurokinin A (also known as substance K [SK], neurokinin alpha, or neuromedin L) nor neurokinin B (also known as neuromedin K [NK] or neurokinin beta) are effective in mimicking the SP effects on the nicotinic response.[238] These results indicate that the SP receptors modulating adrenal catecholamine release are neither of the "SP-P", "SP-E", nor "SP-K" subtype. Rather, bovine adrenal chromaffin cells appear to possess a fourth subtype of tachykinin receptor ("SP-S" receptor, for substance P, the tachykinin most active at this receptor),[211] which shares similar characteristics with the tachykinin receptors mediating SP-induced histamine release from mast cells.[212-214]

b. SP as a Modulator In Vivo

The various observations that SP was able to protect the nicotinic response(s) of chromaffin cells in vitro led to the idea that SP might function in vivo to maintain a prolonged adrenal catecholamine output during times of stress.[7] In support of this hypothesis, we have shown that neonatal pretreatment of rats with capsaicin prevents insulin-stress-induced secretion of adrenaline from the adrenal gland when the rats are stressed as adults.[9,153]

Given the presence of high levels of SP in the splanchnic nerve of several species, the question arises as to whether these SP-containing fibers have any physiological role. Araki et al.[215,216] have shown in the rat that somatic afferent stimulation can influence adrenal secretion. In anesthetized rats, noxious cutaneous stimulation produced reflex increases in sympathetic efferent nerve activity leading to an increased secretion of adrenal catecholamines. In contrast, non-noxious cutaneous stimulation produced reflex decreases in nerve activity and adrenal secretion. They concluded that adrenal medullary secretion of catecholamines could be controlled reflexly via the adrenal sympathetic efferent nerves in either an excitatory or inhibitory way by noxious or nonnoxious mechanical cutaneous stimuli.

Although it is not known whether the viscero-adrenal reflexes produced through stimulation of cutaneous afferents in the rat are conveyed to the adrenal via the SP-containing fibers within the splanchnic nerve, there is evidence that capsaicin-sensitive fibers within the splanchnic nerve play a role in cardiovascular regulation. Stimulation of the splanchnic nerve central to a transection (afferent stimulation) in the rat evokes a reflex fall in blood pressure, but no bradycardia.[152] The fall in blood pressure is absent in rats treated as adults with capsaicin (50 mg/kg), indicating that it is mediated by capsaicin-sensitive primary afferents in the splanchnic nerve.[152] In addition, Lembeck and Skofitsch[217] have shown that adult rats pretreated neonatally with capsaicin had a lowered blood pressure (-30%) and heart rate (-10%). Sprague-Dawley and Wistar rats treated as adults with capsaicin 4 days before measurement also exhibited a reduced mean arterial blood pressure (-16%), although in this case there was no significant decrease in heart rate.

The reflex reduction in blood pressure following activation of the capsaicin-sensitive fibers in the splanchnic nerve did not operate through the baroreceptor reflex, as had been proposed by others,[218,219] but through a reduction in sympathetic vasoconstrictor tone. At present it is not known whether activity within these sensory splanchnic nerve fibers operates normally to evoke similar cardiovascular reflexes and what their physiological significance might be.

In other sensory pathways, capsaicin has been reported to deplete sensory neurones of several neuropeptides in addition to SP, so that it is not clear whether the reflexes observed above were mediated by sensory fibers containing SP or by other classes of capsaicin-sensitive nerves. Recent studies from our own laboratory[239] do, however, support a primary role for the SP-containing neurones in the splanchnic nerve in these reflexes. We have found that neonatal capsaicin pretreatment is very selective in depleting SP from the splanchnic nerve of the rat. Other neuropeptides known to be present in the splanchnic nerve (somatostatin, NPY) were not depleted at the same time that SP was depleted by 70%. We have also confirmed that capsaicin treatment of adult rats, as performed by Lembeck and Donnerer,[152] results in a poorer and more variable depletion of SP-LI from the splanchnic nerve compared to that achieved by neonatal treatment. It would, therefore, be of some interest to repeat these studies on the cardiovascular effects of afferent and efferent splanchnic nerve stimulation, but in rats pretreated with capsaicin as neonates.

Do sensory nerves within the splanchnic nerve have any peripheral functions within the adrenal gland? Sensory nerves in other tissues have been implicated in mechanisms of cutaneous vasodilation and plasma extravasation,[6] however, this appears not to be the case in the adrenal medulla. Experiments in which the peripheral cut end of the splanchnic nerve was stimulated failed to produce plasma extravasation in the adrenal gland of Sprague-Dawley rats.[152]

Recent experiments from our own laboratory show, however, that capsaicin-sensitive sensory afferents and SP do play a role in regulating catecholamine secretion from the adrenal medulla in vivo in response to stress.[9,153] Neonatal capsaicin treatment has a profound effect

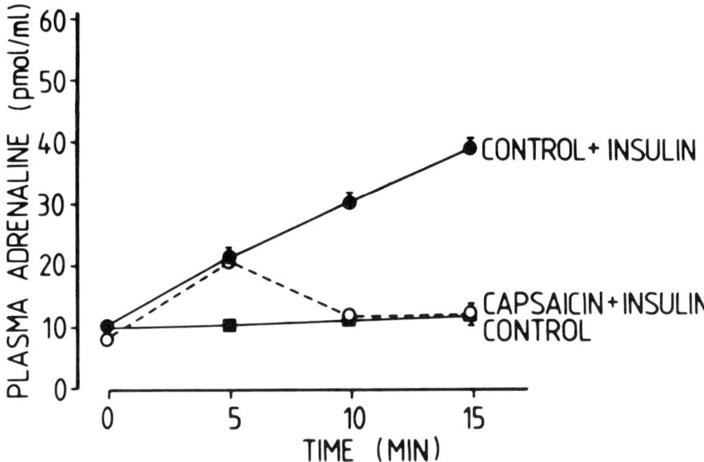

FIGURE 5. Effect of neonatal capsaicin pretreatment of rats on the stress-induced secretion of catecholamines. Plasma adrenaline samples in control rats (vehicle alone, no insulin) and in control (vehicle pretreated) and capsaicin-pretreated rats injected i.v. with insulin (1 IU/kg). Blood samples were withdrawn at 5-min intervals over a 15-min period. All data are plotted as mean ± S.D., n = 5. The plasma noradrenaline levels in the three groups of rats were not significantly different from one another. (From Khalil, Z., Livett, B. G., and Marley, P. D., *J. Physiol. (London)*, 370, 201, 1986. With permission.)

on subsequent insulin-stress-induced catecholamine secretion in vivo in adult Buffalo rats[153] (Figure 5). In vitro, SP protects the chromaffin cells' nicotinic response from desensitization.[8,54,201] In vivo, capsaicin pretreatment depletes SP from the splanchnic nerve and hence removes this protection against desensitization. Consequently, during insulin stress, adrenal catecholamine secretion is abolished since the nicotinic receptor quickly desensitizes.

These results provide physiological relevance to the pharmacological studies in vitro with cultured bovine and guinea pig adrenal chromaffin cells[7,146,147,220] where SP was shown to be a potent modulator of the nicotinic secretion of catecholamines. While other explanations are possible, the idea that SP may be the endogenous component responsible for postsynaptic modulation of the nicotinic response in the adrenal gland is a very appealing one: it also raises the possibility that SP acts similarly at other sites in the autonomic nervous system where ACh and SP interact.

3. Effect of Somatostatin and Analogs on Secretion

Somatostatin, like substance P, causes a noncompetitive inhibition of the nicotinic response in isolated bovine and guinea pig chromaffin cells.[147,198,200,222] Like SP, it exhibits contingent action in its ability to inhibit catecholamine secretion from isolated chromaffin cells in culture, but on a molar basis it is 15-fold less potent than SP.[221] Its potential as a physiological modulator of adrenal catecholamine secretion is made less likely by its localization primarily (if not exclusively) within the chromaffin cells themselves.[123] In addition, somatostatin-like immunoreactivity is present only in very low amounts (e.g., in guinea pig-isolated chromaffin cells approximately 0.7 ± 0.1 pmol/mg protein).[147] Three somatostatin analogs (D-Trp[8]- and D-Trp[8],D-Cys[14]-somatostatin; and des-Asn[5],D-Trp[8],D-ser[13]-somatostatin) known to have greater potency than somatostatin for the somatostatin receptors responsible for inhibition of growth hormone, glucagon, and insulin secretion, respectively, had less potency than somatostatin when tested for their ability to inhibit the nicotinic catecholamine release from cultured bovine adrenal chromaffin cells.[200] A functional role for the somatostatin present in the chromaffin cells has not been established.

4. Effects of Other Neuropeptides on Secretion

A number of other neuropeptides known to be present or have receptors in the adrenal of various species (neurotensin, vasoactive intestinal polypeptide [VIP], neuropeptide Y [NPY], vasopressin, oxytocin, angiotensin II, and bradykinin) have been tested for their ability to modulate catecholamine secretion.[6] Of these, only NPY has been shown to have a direct effect.

Neuropeptide Y (NPY) was originally isolated from porcine brain[223,224] and has been found to be highly concentrated in the brain, heart, and adrenal gland[225-230] in intimate association with sympathetic nerves. Three classes of action have been described for NPY: (1) a direct action on vascular smooth muscle, (2) an inhibition of transmission at autonomic neuroeffector junctions,[6] and (3) an inhibition of basal adrenaline secretion from the adrenal medulla in anesthetized rats with consequent inhibition of the pressor response to noradrenaline and lowering of the blood pressure.[231] This latter study not only indicates a modulatory role for NPY in vivo, but also points to a physiologically important role for the adrenal medulla in the maintenance of blood pressure in the adult rat. In view of the close structural similarities of (porcine) NPY and (bovine) pancreatic polypeptide (BPP) (see Table 7, Marley and Livett[6]), it will be of interest to see whether the late non-neuronal elevation in plasma catecholamines observed in the anesthetized rat in response to insulin hypoglycemia (that is, abolished by pancreatectomy or by administration of glucose)[67,232,233] is a consequence of removal of a similar inhibitory influence of pancreatic polypeptide on the adrenal catecholamine secretion, or due to direct effects of NPY itself on the adrenal.

One peptide that has not been tested to date, but which warrants serious consideration because of its existence in the pancreas and its coexistence with SP in many sites in the peripheral and central nervous system is calcitonin gene related peptide, CGRP.[234] However, given the rate at which new peptides are applied to a variety of systems these days, I imagine that soon the answers will be known.

IV. CONCLUSIONS

This review has considered various forms of control that operate in vitro and in vivo to regulate secretion from the adrenal.

Neuronal control of secretion is obviously important in vivo, though, surprisingly, it appears to contribute less than 30% to the total circulating catecholamines present in the blood 30 to 90 min after a stressful encounter.[67] The nature of the nonneuronal influence contributing to the balance of secretion (70%) is presently a subject of active research interest. Of importance in vivo is the pattern of splanchnic nerve stimulation[31] and the presence of peptides such as substance P that can protect the neuronal response against nicotinic desensitization in times of stress.[9] Developmental considerations are important too as evidenced by the differential sensitivity of the neonate and the adult to opiate-induced secretion.[38] Such non-neuronal endocrine or paracrine influences of endogenous opioid peptides may extend into adult life, but as yet there is no evidence for a functional opioid peptide involvement in adrenal catecholamine secretion. It is likely that some of the neuropeptides present in the adrenal gland and its innervation have regulatory or metabolic functions other than those related to receptor-mediated control of chromaffin cell secretion.

Studies in vitro have encouraged further examination of the finding of diminished secretion upon repeated exposure to agonist, termed "desensitization", and have provided greater insight into this process, though as yet the molecular mechanism(s) operating in tachyphylaxis remains obscure. Isolated adrenal chromaffin cells[7,75] have been most useful in characterizing postsynaptic actions of putative neurotransmitters and neurohormones and distinguishing these actions from presynaptic actions. For example, peptide hormones such as glucagon and bombesin which are active secretagogues when injected in vivo[31] are not active on isolated adrenal chromaffin cells in vitro. The classical adrenoreceptor agonist clonidine is active in vitro, but acts not throught the α receptors, but through inhibition of the nicotinic response.[119,120]

The key questions that arise are (1) what is the balance between these various control processes in vivo, and (2) what is the mechanism by which certain endogeous neuropeptides protect the nicotinic response against neuronal and agonist-induced desensitization? Together the studies reviewed above give a new perspective for control of adrenal medullary secretion and suggest possibilities for the maintenance of homeostasis beyond the simple neuronal release of ACh from splanchnic nerve terminals.

ACKNOWLEDGMENTS

The author wishes to thank Dr. P. D. Marley for his comments on the manuscript, Miss Mafalda Bojanic and Miss Jean Kingett for processing the references, and Mr. Ken I. Mitchelhill for photography. This work was supported by a project grant to B.G.L. from the National Health and Medical Research Council of Australia.

REFERENCES

1. **Lewis, G. P.,** Physiological mechanisms controlling secretory activity of adrenal medulla, in *Handbook of Physiology,* Vol. 6, Sect. 7, Academic Press, New York, 1975, 309.
2. **Feldberg, W., Minz, B., and Tsudzimura, H.,** The mechanism of the nervous discharge of adrenaline, *J. Physiol. (London),* 81, 286, 1934.
3. **Katz, B.,** *The Release of Neural Transmitter Substances,* Liverpool University Press, Liverpool, England, 1969.
4. **Banks, P.,** Effects of stimulation by carbachol on the metabolism of the bovine adrenal medulla, *Biochem. J.,* 97, 555, 1965.
5. **Douglas, W. W. and Rubin, R. P.,** The role of calcium in the secretory response of the adrenal medulla to acetylcholine, *J. Physiol. (London),* 159, 40, 1961.
6. **Marley, P. and Livett, B. G.,** Neuropeptides in the autonomic nervous system, *CRC Crit. Rev. Clin. Neurobiol.,* 1, 201, 1985.
7. **Livett, B. G., Boksa, P., Dean, D. M., Mizobe, F., and Lindenbaum, M. H.,** Use of isolated chromaffin cells to study basic release mechanisms, *J. Auton. Nerv. Syst.,* 7, 59, 1983.
8. **Boksa, P. and Livett, B. G.,** Substance P protects against desensitization of the nicotinic response is isolated adrenal chromaffin cells, *J. Neurochem.,* 42, 618, 1984a.
9. **Khalil, Z., Livett, B. G., and Marley, P. D.,** The role of sensory fibres in the rat splanchnic nerve in the modulation of adrenal medullary secretion during stress, *J. Physiol. (London),* 370, 201, 1986.
10. **Holz, R. W., Senter, R. A., and Frye, R. A.,** Relationship between Ca^{++} uptake and catecholamine secretion in primary dissociation cultures of adrenal medulla, *J. Neurochem.,* 39, 635, 1982.
11. **Kidokoro, Y., Miyazaki, S., and Ozawa, S.,** Acetylcholine-induced membrane depolarization and potential fluctuations in the rat adrenal chromaffin cell, *J. Physiol. (London),* 324, 203, 1982.
12. **Kidokoro, Y. and Ritchie, A. K.,** Chromaffin cell action potential and their possible role in adrenaline secretion from rat adrenal medulla, *J. Physiol. (London),* 307, 199, 1980.
13. **Douglas, W. W., Kanno, T., and Sampson, S. R.,** Effects of acetylcholine and other medullary secret-agogues and antagonists on the membrane potential of adrenal chromaffin cells: an analysis employing techniques of tissue culture, *J. Physiol. (London),* 188, 107, 1967a.
14. **Douglas, W. W., Kanno, T., and Sampson, S. R.,** Influence of the ionic environment on the membrane potential of adrenal chromaffin cells and on the depolarizing effect of acetylcholine, *J. Physiol. (London),* 191, 107, 1967b.
15. **Kanno, T. and Douglas, W. W.,** Effect of rapid application of acetylcholine or depolarizing current on transmembrane potentials of adrenal chromaffin cells, *Proc. Can. Fed. Biol. Soc.,* 10, 39, 1967.
16. **Matthews, E. K.,** Membrane potential measurement in cells of the adrenal gland, *J. Physiol. (London),* 189, 139, 1967.
17. **Baird, A. R., Benoit, R., Ling, A., Bohlen, P., and Guillemin, R.,** Partial characterization of soma-tostatin-like material in the bovine adrenal medulla, *Soc. Neurosci. Abstr.,* 7, 606, 1981.

18. **Brandt, B. L., Hagiwara, S., Kidokoro, Y., and Miyazaki, S.,** Action potentials in the rat chromaffin cell and effects of acetylcholine, *J. Physiol. (London)*, 263, 417, 1976.

19. **Kidokoro, Y. and Ritchie, A. K.,** Chromaffin cells action potential and their possible role in adrenaline secretion from rat adrenal medulla, *J. Physiol. (London)*, 307, 199, 1980.

20. **Maruyama, T.,** Patterns of membrane excitability in developing chick adrenal chromaffin cells in primary culture, *Biomed. Res.*, 1, 112, 1980.

21. **Kehoe, J. S. and Marty, A.,** Certain slow synaptic responses: their properties and possible underlying mechanisms, *Annu. Rev. Biophys. Bioeng.*, 9, 437, 1980.

22. **Marty, A.,** Ca^{2+}-dependent K^+ channels with large unitary conductance, *TINS*, 6, 262, 1983.

23. **Marty, A.,** Ca-dependent K channels with large unitary conductance in chromaffin cell membranes, *Nature (London)*, 291, 497, 1981.

24. **Fenwick, E. M., Fajdiga, P. B., Howe, N. B. S., and Livett, B. G.,** Functional and morphological characterization of isolated bovine adrenal medullary cells, *J. Cell Biol.*, 76, 12, 1978.

25. **Fenwick, E. M., Marty, A., and Neher, E.,** Voltage clamp and single channel recording from bovine chromaffin cells, *J. Physiol. (London)*, 331, 577, 1982.

26. **Fenwick, E. M., Marty, A., and Neher, E.,** A patch clamp study of bovine chromaffin cells, *J. Physiol. (London)*, 331, 599, 1982.

27. **Kesteven, N. T. and Knight, D. E.,** Transient changes of intracellular free Ca associated with catecholamine secretion in isolated bovine adrenal medullary cells, *J. Physiol. (London)*, 328, 57P, 1982.

28. **Knight, D. E. and Kesteven, N. T.,** Evoked transient intracellular Ca^{2+} changes and secretion in isolated bovine adrenal medullary cells, *Proc. R. Soc. London B*, 218, 177, 1983.

29. **Kirpekar, S. M., Prat, J. C., and Schiavone, M. T.,** Modification of potassium-evoked release of noradrenaline by various ions and agents, *Br. J. Pharmacol.*, 78, 277, 1983.

30. **Kanner, B., Fishkes, H., Maron, R., Sharon, I., and Shulinder, S.,** Reserpine as a competitive and reversible inhibitor of the catecholamine transporter of bovine chromaffin granules, *FEBS Lett.*, 100, 175, 1979.

31. **Edwards, A. V., Bircham, P. M. M., O'Connor, B. N., Ghatei, M. A., McGregor, G. P., Adrian, T. E., and Bloom, S. R.,** The importance of the pattern of activity in sympathetic nerves, in *Catecholamines: Basic and Peripheral Mechanisms*, Usdin, E., Carlsson, A., Dahlstrom, A., and Engel, J., Eds., Alan R. Liss, New York, 1984, 129.

32. **Comline, R. S. and Silver, M.,** The development of the adrenal medulla of the foetal and new-born calf, *J. Physiol. (London)*, 183, 385, 1966.

33. **Silver, M.,** The output of adrenaline and noradrenaline from the adrenal medulla of the calf, *J. Physiol. (London)*, 152, 14, 1980.

34. **Lemaire, S., Dumont, M., Mercier, P., Lemaire, I., and Calvert, R.,** Biochemical characterization of various populations of isolated bovine adrenal cells, *Neurochem. Int.*, 5, 193, 1983.

35. **Edwards, A. V., Jarhult, J., Andersson, P.-O., and Bloom, S. R.,** The importance of the pattern of stimulation in relation to the response of autonomic effectors, in *Systemic Role of Regulatory Peptides*, Bloom, S. R., Polak, J. M., and Lindenlaub, P., Eds., Schattauer-Verlag, Stuttgart, 1983, 145.

36. **Slotkin, T. A., Smith, P. G., Lau, C., and Bareis, D. L.,** Functional aspects of development of catecholamine biosynthesis and release in the sympathetic nervous system, in *Biogenic Amines in Development*, Parvez, H. and Parvez, S., Eds., Elsevier/North-Holland, Amsterdam, 1980, 29.

37. **Anderson, T. R. and Slotkin, T. A.,** The role of neural input in the effects of morphine on the rat adrenal medulla, *Biochem. Pharmacol.*, 25, 1071, 1976.

38. **Chantry, C. J., Seidler, F. J., and Slotkin, T. A.,** Non-neurogenic mechanism for reserpine-induced release of catecholamines from the adrenal medulla of neonatal rats: possible modulation by opiate receptors, *Neuroscience*, 7, 673, 1982.

39. **Coupland, R. E.,** The development and fate of catecholamine secreting endocrine cells, in *Biogenic Amines in Development*, Parvez, H. and Parvez, S., Eds., Elsevier/North-Holland, Amsterdam, 1980, 3.

40. **Jones, C. T.,** Circulating catecholamines in the fetus, their origin, actions and significance, in *Biogenic Amines in Development*, Parvez, H. and Parvez, S., Eds., Elsevier/North-Holland, Amsterdam, 1980, 3.

41. **Silver, M. and Edwards, A. V.,** The development of the sympathoadrenal system with an assessment of the role of the adrenal medulla in the foetus and neonate, in *Biocenic Amines in Development*, Parvez, H. and Parvez, S., Eds., Elsevier/North-Holland, Amsterdam, 1980.

42. **Pappano, A. J.,** Ontogenic development of autonomic neuroeffector transmission and transmitter reactivity in embryonic and fetal hearts, *Pharmacol. Rev.*, 29, 3, 1977.

43. **Jones, C. T., Roebuck, M. M., and Walker, D. W.,** The effects of adrenal medullary activity on the sensitivity to catecholamine stimulation of the sympathetic system in fetal sheep, in *Catecholamines: Basic and Peripheral Mechanisms*, Usdin, E., Carlsson, A., Dahlstrom, A., and Engel, J., Eds., Alan R. Liss, New York, 1984, 121.

44. **Celander, O.,** The range of control exercised by sympathico-adrenal system, *Acta Physiol. Scand. Suppl.*, 116, 1, 1954.

45. **Neil, E.,** Catecholamines and the cardiovascular system, in *American Handbook of Physiology,* Vol. 6, Sect. 7, Academic Press, New York, 1975, 473.
46. **Bartolome, J. and Slotkin, T. A.,** Effects of postnatal reserpine administration on sympatho-adrenal development in the rat, *Biochem. Pharmacol.,* 25, 1513, 1976.
47. **Slotkin, T. A.,** Maturation of the adrenal medulla. II. Content and properties of catecholamine storage vesicles of the rat, *Biochem. Pharmacol.,* 22, 2033, 1973.
48. **Lagercrantz, H. and Bistoletti, P.,** Catecholamine release in the newborn infant at birth, *Pediatr. Res.,* 11, 889, 1973.
49. **Yoshizaki, T.,** Participation of muscarinic receptors on splanchnic adrenal transmission in the rat, *Jpn. J. Pharmacol.,* 23, 813, 1973.
50. **Elliott, T. R.,** The control of the suprarenal glands by the splanchnic nerves, *J. Physiol. (London),* 44, 374, 1912.
51. **Costa, E., Guidotti, A., Hanbauer, I., Hexum, T., Saiani, L., and Yang, H.-Y. T.,** Regulation of cholinergic transmission in adrenal medulla, in *Cholinergic Mechanisms,* Papau, G. and Laduisky, H., Eds., Plenum Press, New York, 1981, 143.
52. **Schultzberg, M., Lundberg, J. M., Hökfelt, T., Terenius, L., Brandt, J., Elde, R. P., and Goldstein, M.,** Enkephalin-like immunoreactivity in gland cells and nerve terminals of the adrenal medulla, *Neuroscience,* 3, 1169, 1978.
53. **Dean, D. M. and Livett, B. G.,** Study of mechanism(s) of peptide modulation of catecholamine release in bovine adrenal paraneurons, *Soc. Neurosci. Abstr.,* 6, 335, 1980.
54. **Livett, B. G. and Boksa, P. B.,** Receptors and receptor modulation in cultured chromaffin cells, *Can. J. Physiol. Pharmacol.,* 62, 467, 1984.
55. **Mizuno, K., Minamino, N., Kangawa, K., and Matsuo, H.,** A new family of endogenous "big" met-enkephalins from bovine adrenal medulla: purification and structure of docosa-(bam-22p) and eicosapeptide (bam-20p) with very potent opiate activity, *Biochem. Biophys. Res. Commun.,* 97, 1283, 1980.
56. **Marley, P., Mitchelhill, K. I., and Livett, B. G.,** Metorphamide, a novel endogenous adrenal opioid peptide, inhibits nicotine-induced secretion from bovine adrenal chromaffin cells, *Brain Res.,* 363, 10, 1986.
57. **Lewis, R. V., Stern, A. S., Kilpatrick, D. L., Gerber, L. D., Rossier, J., Stein, S., and Udenfriend, S.,** Marked increases in large enkephalin-containing polypeptides in the rat adrenal gland following denervation, *J. Neurochem.,* 1, 80, 1981.
58. **Viveros, O. H. and Wilson, S. P.,** The adrenal chromaffin cell as a model to study the co-secretion of enkephalins and catecholamines, *J. Auton. Nerv. Syst.,* 7, 41, 1983.
59. **Wilson, S. P., Abou-Donia, M. M., Chang, K.-J., and Viveros, O. H.,** Reserpine increases opiate-like peptide content and tyrosine hydroxylase activity in adrenal medullary chromaffin cells in culture, *Neuroscience,* 6, 71, 1981.
60. **Wilson, S. P., Chang, K.-J., and Viveros, O. H.,** Synthesis of enkephalins by adrenal medullary chromaffin cells: reserpine increases incorporation of radiolabeled amino acids, *Proc. Natl. Acad. Sci. U.S.A.,* 77, 4364, 1980.
61. **Garber, A. J., Cryer, P. E., Santiago, J. V., Haymond, M. W., Pagliara, A. S., and Kipnis, D. M.,** The role of adrenergic mechanisms in the substrate and hormonal response to insulin-induced hypoglycaemia in man, *J. Clin. Invest.,* 58, 7, 1976.
62. **Viveros, O. H. and Unsworth, C. D.,** Regulation of opioid peptide levels in the adrenal medulla: the role of splanchnic innervation, neurogenic stimulation and endogenous catecholamines, in Proc. Int. Symp. on the Molecular Biology of Peripheral Catecholamine Storing Tissues, Colmar, August 5 to 9, 1984, B52.
63. **Comline, R. S. and Silver, M.,** The release of adrenaline and noradrenaline from the adrenal glands of the foetal sheep, *J. Physiol. (London),* 156, 424, 1961.
64. **Cantu, R. C., Wise, B. L., Goldfien, A., Gilluxson, K. S., Fischer, N., and Ganong, W. F.,** Neural pathways mediating the increase in adrenal medullary secretion produced by hypoglycaemia, *Proc. Soc. Exp. Biol. Med.,* 114, 10, 1963.
65. **Khalil, Z., Marley, P. D., and Livett, B. G.,** Insulin-stress induced catecholamine secretion from the adrenal gland in-vivo: neurogenic and non-neurogenic components, *Neurosci. Lett.,* 19(Suppl.), 578, 1985.
66. **Sarcione, E. J., Back, N., Sakol, J. E., Mehlman, B., and Knoblock, E.,** Elevation of plasma epinephrine levels produced by glucagon in-vivo, *Endocrinology,* 72, 523, 1963.
67. **Khalil, Z., Marley, P. D., and Livett, B. G.,** Elevation in plasma catecholamines in response to insulin-stress is under both neuronal and non-neuronal control, *Endocrinology,* 119, 159, 1986.
68. **Douglas, W. W.,** Stimulus-secretion coupling: the concept and clues from chromaffin and other cells, *Br. J. Pharmacol.,* 34, 451, 1968.
69. **Schumann, H. J. and Philippu, A.,** Der Einfluss von Calcium auf die Brenzcatechinaminfreisetzung, *Experientia,* 18, 138, 1962.

70. **Blaschko, H., Comline, R. L., Schneider, F. H., Silver, M., and Smith, A. D.,** Secretion of a chromaffin granule protein, Chromogranin, from the adrenal gland after splanchnic stimulation, *Nature (London),* 215, 58, 1967.

71. **Baker, P. F. and Rink, T. J.,** Catecholamine release from bovine adrenal medulla in response to maintained depolarizations, *J. Physiol. (London),* 253, 593, 1975.

72. **Livett, B. G., Mitchelhill, K. I., and Dean, D. M.,** Adrenal chromaffin cells — their isolation and culture, in *The Secretory Process,* Vol. 3, Poisner, A. and Trifaró, J. M., Eds., Elsevier/North-Holland, Amsterdam, 1987, 171.

73. **Livett, B. G.,** Chromaffin cells and chromaffin granules, in *Encyclopaedia of Neuroscience,* Adelman, G., Ed., Birkhauser Boston, Cambridge, Mass., 1987.

74. **Livett, B. G.,** The secretory process in adrenal medullary cells, in *Cell Biology of the Secretory Process,* Cantin, M., Ed., S. Karger, Basel, 1984, 309.

75. **Livett, B. G.,** Adrenal medullary chromaffin cells in-vitro, *Physiol. Rev.,* 64, 1103, 1984a.

76. **Finer-Moore, J. and Stroud, R. M.,** Amphipathic analysis and possible formation of the ion channel in an acetylcholine receptor, *Proc. Natl. Acad. Sci. U.S.A.,* 81, 155, 1984.

76a. **Fricker, L. D.,** Neuropeptide biosynthesis: focus on the carboxypeptidase processing enzyme, *TINS,* May, 210, 1985.

77. **Mishina, M., Tobimatsu, T., Imoto, K., Tanaka, K., Fujita, Y., Fukuda, K., Kurasaki, M., Takahashi, H., Morimoto, Y., Hirose, T., Inayama, S., Takahashi, T., Kuno, M., and Numa, S.,** Location of functional regions of acetylcholine receptor a-subunit by site-directed mutagenesis, *Nature (London),* 313, 364, 1985.

78. **Quik, M. and Trifaró, J. M.,** The α-bungarotoxin site and its relation to the cholinergic and nerve growth factor mediate increases in tyrosine hydroxylase activity in cultures of sympathetic ganglia and chromaffin cells, *Brain Res.,* 224, 332, 1982.

79. **Boksa, P. and Livett, B. G.,** Desensitization to nicotinic cholinergic agonists and K^+, agents that stimulate catecholamine secretion, in isolated adrenal chromaffin cells, *J. Neurochem.,* 42, 607, 1984a.

80. **Oka, M., Isosaki, J., Watanabe, H., Houchi, H., Tsunematsu, T., and Minaguchi, K.,** Nicotinic and muscarinic receptors in chromaffin cells: their role in the regulation of Ca^{++} flux and catecholamine release and synthesis, in Int. Conf. Molecular Neurobiology of Peripheral Catecholaminergic Systems, Ibiza, Spain, 1982, 70.

81. **Knight, D. E. and Kesteven, N. T.,** Evoked transient intracellular Ca^{2+} changes and secretion in isolated bovine adrenal medullary cells, *Proc. R. Soc. London Ser. B,* B218, 177, 1983.

82. **Brooks, J. C.,** The isolated bovine adrenomedullary chromaffin cell: a possible model for neuronal function (abstract), *Fed. Proc.,* 36, 364, 1977.

83. **Brooks, J. C. and Treml, S.,** Trifluoperazine-induced inhibition of catecholamine secretion by isolated adrenal chromaffin cells, *Soc. Neurosci. Abstr.,* 8, 544, 1982.

84. **Brooks, J. C. and Treml, S.,** Catecholamine secretion by saponin-skinned cultured chromaffin cells, *J. Neurochem.,* 40, 468, 1983.

85. **Knight, D. E.,** Temperature sensitivity of catecholamine release in response to different secretagogues, *J. Physiol. (London),* 298, 41P, 1980.

86. **Chaminade, M., Foutz, A. S., and Rossier, J.,** Co-release of enkephalins and precursors with catecholamines from the perfused cat adrenal gland in-situ, *J. Physiol. (London),* 353, 157, 1984.

87. **Marley, P., Mitchelhill, K. I., and Livett, B. G.,** Adrenal chromaffin cell secretion in response to repeated exposure to stimuli, *Proc. Aust. Endocrinol. Soc.,* 28, 28, 1985c.

88. **Manthey, A. A.,** The effect of calcium on the desensitization of membrane receptors at the neuromuscular junction, *J. Gen. Physiol.,* 49, 963, 1966.

89. **Nastuk, W. L. and Parsons, R. L.,** Factors in the inactivation of postjunctional membrane receptors of frog skeletal muscle, *J. Gen. Physiol.,* 56, 218, 1970.

90. **Katz, B. and Thesleff, S.,** A study of the "desensitization" produced by acetylcholine at the motor-endplate, *J. Physiol. (London),* 138, 63, 1957.

91. **Douglas, W. W. and Rubin, R. P.,** Stimulant action of barium on the adrenal medulla, *Nature (London),* 203, 305, 1964.

92. **Bevington, A. and Radda, G. K.,** Enhanced oxygen consumption in adrenal medulla on stimulation with acetylcholine, *Biochem. Pharmacol.,* 34, 211, 1985.

93. **Itoh, S., Nakazato, Y., and Ohga, A.,** Pharmacological evidence for the involvement of Na^+ channels in the release of catecholamines from perfused adrenal glands, *Br. J. Pharmacol.,* 62, 359, 1978.

94. **Magazanik, L. G. and Viskocil, F.,** The effect of temperature on desensitization kinetics at the postsynaptic membrane of the frog muscle fibre, *J. Physiol. (London),* 249, 285, 1975.

95. **Burgermeister, W., Catterall, W. A., and Witkop, B.,** Histrioicotoxin enhances agonist-induced desensitization of acetylcholine receptor, *Proc. Natl. Acad. Sci. U.S.A.,* 74, 5754, 1977.

96. **Stallcup, W. B. and Patrick, J.,** Substance P enhances cholinergic receptor desensitization in a clonal nerve cell line, *Proc. Natl. Acad. Sci. U.S.A.,* 77, 634, 1980.

97. **Boksa, P.,** Effects of substance P on carbachol-stimulated $^{45}Ca^{2+}$ uptake into cultures adrenal chromaffin cells, *J. Neurochem.,* 45, 1895, 1985.
98. **Schiavone, M. T. and Kirpekar, S. M.,** Inactivation of secretory responses to potassium and nicotine in the cat adrenal medulla, *J. Pharmacol. Exp. Ther.,* 223, 743, 1982.
99. **Bevington, A. and Radda, G. K.,** Declining catecholamine secretion in adrenal medulla on prolonged stimulation with acetylcholine, *Biochem. Pharmacol.,* 34, 1497, 1985.
100. **Garcia, A. G., Hernandez, M., Horga, J. F., and Sanchez-Garcia, P.,** On the release of catecholamines and dopamine B-hydroxylase evoked oubain in the perfused cat adrenal gland, *Br. J. Pharmacol.,* 68, 571, 1980.
101. **Ledbetter, F. H. and Kirshner, N.,** Studies of chick adrenal medulla in organ culture, *Biochem. Pharmacol.,* 24, 967, 1975.
102. **Ledbetter, F. H. and Kirshner, N.,** Quantitative correlation between secretion and cellular content of catecholamines and dopamine β-hydroxylase in cultures of adrenal medulla cells, *Biochem. Pharmacol.,* 30, 3246, 1981.
103. **Derome, G., Tseng, R., Mercier, P., Lemaire, I., and Lemaire, S.,** Possible muscarinic regulation of catecholamine secretion mediated by cyclic gmp in isolated bovine adrenal chromaffin cells, *Biochem. Pharmacol.,* 30, 855, 1981.
104. **Fisher, S. K., Holz, R. W., and Agranoff, B. W.,** Muscarinic receptors in chromaffin cell cultures enhanced phospholipid labelling but not catecholamine secretion, *J. Neurochem.,* 37, 491, 1981.
105. **Lemaire, S., Denis, D., and Day, R.,** Dynorphin in bovine adrenal medulla: isolation, identification and secretion, *Proc. 8th Int. Soc. Neurochem.,* p. 209, 1981.
106. **Schneider, A. S., Cline, H. T., and Lemaire, S.,** Rapid rise in cyclic GMP accompanies catecholamine secretion in suspensions of adrenal chromaffin cells, *Life Sci.,* 24, 1389, 1979.
107. **Yanagihara, N., Isosaki, M., Ohuchi, T., and Oka, M.,** Muscarinic receptor-mediated increase in GMP level in isolated bovine adrenal medullary cells, *FEBS Lett.,* 105, 296, 1979.
108. **Berridge, M. J.,** Inositol trisphosphate and diacyglycerol as second messengers, *Biochem. J.,* 220, 345, 1984.
109. **Gutman, Y. and Boonyaviroj, P.,** Suppression by noradrenaline of catecholamine secretion from the adrenal medulla, *Eur. J. Pharmacol.,* 28, 384, 1974.
110. **Gutman, Y. and Boonyaviroj, P.,** Inhibition of catecholamine release by alpha-adrenergic activation: interaction with Na,K-ATPase, *J. Neural Transmission,* 40, 245, 1977.
111. **Gutman, Y.,** The mechanisms of PGE, alpha-adrenergic and beta-adrenergic regulation of catecholamine release: evidence for different modes of action, in *Catecholamines: Basic and Clinical Frontiers.* Vol. 1, Usdin, E., Kopin, I. J., and Barchas, J., Eds., Pergamon Press, New York, 1979, 277.
112. **Greenberg, A. and Zinder, O.,** Alpha- and β-receptor control of catecholamine secretion from isolated adrenal medulla cells, *Cell Tissue Res.,* 226, 655, 1982.
113. **Wada, A., Sakurai, H., Kobayashi, H., Yanagihara, N., and Izumi, F.,** Alpha 2-adrenergic receptors inhibit catecholamine secretion from bovine adrenal medulla, *Brain Res.,* 252, 189, 1982.
114. **Sakurai, S., Wada, A., Izumi, F., Kobayashi, H., and Yanagihara, N.,** Inhibition by alpha 2-adrenergic agonists of the secretion of catecholamines from isolated adrenal medullary cells, *Naunyn-Schmiedeberg's Arch. Pharmacol.,* 324, 15, 1983.
115. **Starke, K.,** Regulation of noradrenaline release by presynaptic receptor systems, *Rev. Physiol. Biochem. Pharmacol.,* 77, 1, 1977.
116. **Collett, A. R. and Story, D. F.,** Release of H^3-adrenaline from an isolated intact preparation of the rabbit adrenal gland: no evidence for release modulatory alpha-adrenoreceptors, *J. Auton. Pharmacol.,* 2, 25, 1982.
117. **Collett, A. R., Rand, M. J., and Story, D. F.,** The effects of alpha 2-adrenoceptor agonists and antagonists on the release of H^3-adrenaline from a rabbit isolated adrenal preparation, *Clin. Exp. Pharmacol. Physiol.,* 8, 611, 1982.
118. **Ladona, M. G., Montiel, C., Ceña, V., Artalejo, A. R., Sanchez-Garcia, P., and Garcia, A. G.,** Modulation of adrenomedullary catecholamine release by adrenoceptors, dopaminoceptors and cholinoceptors, in Molecular Biology of Peripheral Catecholamine Storing Tissues, IUPHAR Satellite Meeting, Abstr. HP4, Colmar, France, August 5 to 9, 1984, 132.
119. **Powis, D. A.,** Clonidine inhibits catecholamine release from bovine adrenal medullary cells but apparently not via $alpha_2$-adrenoceptors, *J. Physiol. (London),* 358, 87P, 1984.
120. **Powis, D. A. and Baker, P. F.,** Alpha 2-adrenoceptors do not regulate catecholamine secretion by bovine adrenal medullary cells: a study with clonidine, *Mol. Pharmacol.,* 29, 134, 1986.
121. **Staszewska-Barcak, J. and Vane, J. R.,** The release of catecholamines from the adrenal medulla by peptides, *Br. J. Pharmacol.,* 30, 665, 1967.
122. **Helle, K.,** Some chemical and physical properties of the soluble protein fraction of bovine adrenal chromaffin granules, *Mol. Pharmacol.,* 2, 298, 1966.

123. **Linnoila, R. I., Diaugustine, R. P., Hervonen, A., and Miller, R. J.,** Distribuation of (Met5)- and (Leu5)-enkephalin, vasoactive-intestinal polypeptide, and substance P-like immunoreactivities in human adrenal glands, *Neuroscience,* 5, 2247, 1980.

124. **Hökfelt, T., Johansson, O., Ljungdahl, A., Lundberg, J. M., and Schultzberg, M.,** Peptidergic neurones, *Nature (London),* 284, 515, 1980.

125. **Livett, B. G., Day, R., Elde, R. P., and Howe, P. R. C.,** Co-storage of enkephalins and adrenaline in the bovine adrenal medulla, *Neuroscience,* 7, 1323, 1982.

126. **Livett, B. G. and Dean, D. M.,** Distribution of immunoreactive enkephalins in adrenal paraneurons: preferential localization in varicose processes and terminals, *Neuropeptides,* 1, 3, 1980.

127. **Livett, B. G., Day, R., Howe, P. R. C., and Elde, R. P.,** Co-storage and leu- and met-enkephalin within adrenaline synthesizing cells in the bovine adrenal medulla, *Proc. 8th Int. Soc. Neurochem. (Nottingham, U.K.),* p. 160, 1981.

128. **Tischler, A. S., De Lellis, R. A., Slayton, V. W., Blount, M. W., and Wolfe, H. J.,** Enkephalin-like immunoreactivity in human adrenal medullary cultures, *Lab. Invest.,* in press.

129. **Livett, B. G., Dean, D. M., Whelan, L. G., Udenfriend, S., and Rossier, J.,** Co-release of leu-enkephalin and catecholamines from adrenal chromaffin cells in culture, *Nature (London),* 289, 317, 1981.

130. **Rosin, M. P., Artola, A., Henry, J. P., and Rossier, J.,** Enkephalins are associated with adrenergic granules in bovine adrenal medulla, *Neuroscience,* 10, 83, 1983.

131. **Lang, R. E., Taugner, G., Gaida, W., Ganten, D., Kraft, K., Unger, T., and Wunderlich, I.,** Evidence against co-storage of enkephalins with noradrenaline in bovine adrenal medullary granules, *Eur. J. Pharmacol.,* 86, 117, 1982.

132a. **Varndell, I. M., Tapia, F. J., Demey, J., Rush, R. A., Bloom, S. R., and Polak, J. M.,** Electroimmunocytochemical localization of enkephalin-like material in phaeochromocytomas of man and other animals, *J. Histochem. Cytochem.,* 30, 682, 1982.

132b. **Kobayashi, S., Ohashi, T., Fujita, T., Nakao, K., Yoshimasa, T., Imura, H., Mochizuki, T., Yanaihara, C., Yanaihara, N., and Verhofstad, A. A. J.,** An immunohistochemical study on co-storage of met-enkephalin-arg-leu and met-enkephalin-arg-phe with adrenaline and/or noradrenaline in the adrenal chromaffin cells of the rat, dog and cat, *Biomed. Res.,* 4, 433, 1983.

132c. **Bloch, B., Milner, R. J., Baird, A., Gubler, U., Reymond, C., Bohlen, P., le Guellac, D., and Bloom, F. E.,** Detection of the messenger RNA coding for preproenkephalin A in bovine adrenal by in-situ hybridization, *Reg. Peptides,* 8, 345, 1984.

133. **Lemaire, S., Chouinard, L., Denis, D., Panico, M., and Morris, H. R.,** Mass spectrometric identification of various molecular forms of dynorphin in bovine adrenal medulla, *Biochem. Biophys. Res. Commun.,* 108, 51, 1982.

134. **Day, R., Denis, D., Barabe, J., St. Pierre, S., and Lemaire, S.,** Dynorphin in bovine adrenal medulla. I. Detection in glandular and cellular extracts and secretion from isolated chromaffin cells, *Int. J. Peptide Res.,* 19, 10, 1982.

135. **Watson, S. J., Akil, H., Ghazarossian, V. E., and Goldstein, A.,** Dynorphin immunocytochemical localisation in brain and peripheral nervous system: preliminary studies, *Proc. Natl. Acad. Sci. U.S.A.,* 78, 1260, 1981.

136. **Lemaire, S., Day, R., Dumont, M., Chouinard, L., and Calvert, R.,** Dynorphin and enkephalins in adrenal paraneurones. Opiates in the adrenal medulla, *Can. J. Physiol. Pharmacol.,* 62, 484, 1984.

137. **Tan, L. and Wu, P. H.,** Biosynthesis of enkephalins by chromaffin cells of bovine adrenal medulla, *Biochem. Biophys. Res. Commun.,* 95, 1901, 1980.

138. **Denis, D., Day, R., and Lemaire, S.,** Dynorphin in bovine adrenal medulla, *Int. J. Peptide Res.,* 19, 18, 1982.

139a. **Dumont, M., Day, R., and Lemaire, S.,** Distinct distribution of immunoreactive dynorphin and leucine enkephalin in various populations of isolated adrenal chromaffin cells, *Life Sci.,* 32, 287, 1983.

139b. **Pelto-Huikko, M., Salminen, T., and Hervonen, A.,** Enkephalin-like immunoreactivity is restricted to the adrenaline cells in the hamster adrenal medulla, *Histochemistry,* 73, 493, 1982.

140. **Gamse, R., Wax, Z., Zigmond, R. E., and Leeman, S. E.,** Immunoreactive substance P in sympathetic ganglia: distribution and sensitivity towards capsaicin, *Neuroscience,* 6, 437, 1981.

141. **Brodin, E. and Nilsson, G.,** Concentration of substance P-like immunoreactivity in tissues of dog, rat and mouse, *Acta Physiol. Scand.,* 112, 305, 1981.

142. **Bucsics, A., Saria, A., and Lembeck, F.,** Substance P in the adrenal gland: origin and species distribution, *Neuropeptides,* 1, 329, 1981.

143. **Saria, A., Wilson, S. P., Molnar, A., Viveros, O. H., and Lembeck, F.,** Substance P and opiate-like peptides in human adrenal medulla, *Neurosci. Lett.,* 20, 194, 1980.

144. **Saria, A., Molnar, A., and Lembeck, F.,** Immunoreactive substance P in the adrenal medulla, *Arch. Pharmacol.,* 313(Suppl. R24, Abstr. 96), 1980a.

145. **Nilsson, G. and Brodin, E.,** Tissue distribution of substance P-like immunoreactivity in dog, cat, rat and mouse, in *Substance P,* Von Euler, U. S. and Pernow, B., Eds., Raven Press, New York, 1977, 49.

146. **Role, L. W., Perlman, R. L., and Leeman, S. E.,** Somatostatin and substance P inhibit catecholamine secretion from guinea pig chromaffin cells, *Soc. Neurosci. Abstr.,* 5, 597, 1979.

147a. **Role, L. W., Leeman, S. E., and Perlman, R. L.,** Somatostatin and substance P inhibit catecholamine secretion from isolated cells of guinea-pig adrenal medulla, *Neuroscience,* 6, 1813, 1981.

147b. **Scian, L. F., Westerman, C. D., Verdesca, A. S., and Hilton, J. G.,** Adrenocortical and medullary effects of glucagon, *Am. J. Physiol.,* 199, 867, 1960.

148. **Pfister, V. C. and Görne, R. C.,** Substance-P-like immunofluorescence in the adrenal medulla of the rat, *Acta Histochem.,* 72, 127, 1983.

149. **Dalsgaard, C.-J., Hökfelt, T., Elfvin, L.-G., Skirboll, L., and Emson, P. C.,** Substance P-containing primary sensory neurons projecting to the inferior mesenteric ganglion: evidence from combined retrograde tracing and immunohistochemistry, *Neuroscience,* 7, 647, 1982.

150. **Matthews, M. R. and Cuello, A. C.,** Substance P immunoreactive peripheral branches of sensory neurons innervate guinea pig sympathetic neurons, *Proc. Natl. Acad. Sci. U.S.A.,* 79, 1668, 1982.

151. **Holets, V. and Elde, R.,** Sympathoadrenal pre-ganglionic neurons: their distribution and relationship to chemically-coded fibres in the kitten intermediolateral cell column, *J. Auton. Nerv. Syst.,* 7, 149, 1983.

152. **Lembeck, F. and Donnerer, J.,** Reflex fall in blood pressure mediated by capsaicin-sensitive afferent fibers of the rat splanchnic nerve, *Naunyn Schmiedeberg's Arch. Pharmacol.,* 322, 268, 1983.

153. **Khalil, Z., Marley, P. D., and Livett, B. G.,** Neonatal capsaicin treatment prevents insulin-stress-induced adrenal catecholamine secretion in vivo: possible involvement of sensory nerves containing substance P, *Neurosci. Lett.,* 45, 65, 1983.

154. **Dean, D. M., Lemaire, S., and Livett, B. G.,** Evidence that inhibition of nicotine-mediated catecholamine secretion from adrenal chromaffin cells by enkephalin, β-endorphin, dynorphin (1-13), and opiates is not mediated via specific opiate receptors, *J. Neurochem.,* 38, 606, 1982.

155. **Marley, P., Mitchelhill, K. I., and Livett, B. G.,** The effects of opioid peptides containing the sequence of met^{5}enkephalin or leu^{5}enkephalin on nicotine-induced secretion from bovine adrenal chromaffin cells, *J. Neurochem.,* 46, 1, 1986.

156. **Lewis, R. V. and Stern, A. S.,** Biosynthesis of the enkephalins and enkephalin-containing polypeptides, *Ann. Rev. Pharmacol. Toxicol.,* 23, 353, 1983.

157. **Udenfriend, S. and Kilpatrick, D. L.,** Biochemistry of the enkephalins and enkephalin-containing peptides, *Arch. Biochem. Biophys.,* 221, 309, 1983.

158. **Rossier, J., Liston, D., Patey, G., Chaminade, M., Foutz, A. S., Cupo, A., Giraud, P., Roisin, M. P., Henry, J. P., Verbank, P., and Vanderhagen, J.-J.,** The enkephalinergic neuron: implications of a polyenkephalin precursor, *Cold Spring Harbor Symp. Quant. Biol.,* 68, 393, 1983.

159. **Kojima, K., Kilpatrick, D. L., Stern, A. S., Jones, B. N., and Udenfriend, S.,** Proenkephalin: a general pathway for enkephalin biosynthesis in animal tissues, *Arch. Biochem. Biophys.,* 215, 638, 1982.

160. **Fricker, L.-D.,** Neuropeptide biosynthesis: focus on the carboxypeptidase processing enzyme, *TINS,* May, 210, 1985.

161. **Liston, D., Patey, G., Rossier, J., Verbanck, P., and Vanderhaegen, J.-J.,** Processing of proenkephalin is tissue-specific, *Science,* 225, 734, 1984.

162. **Giraud, A. S., Williams, R. G., and Dockray, G. J.,** Evidence for different patterns of post-translational processing of proenkephalin in the bovine adrenal, colon, and striatum indicated by radioimmunoassay using region-specific antisera to met-enk-arg^6-Phe^7 and met-enk-arg^6-gly^7-leu^8, *Neurosci. Lett.,* 46, 223, 1984.

163. **Beaumont, A., Metters, K. M., Rossier, J., and Hughes, J.,** Identification of a proenkephalin precursor in striatal tissue, *J. Neurochem.,* 44, 934, 1985.

164. **Patey, G., Liston, D., and Rossier, J.,** Characterization of new enkephalin-containing peptides in the adrenal medulla by imunoblotting, *FEBS Lett.,* 172, 303, 1984.

165. **Prado, E. S., Prado De Carvalho, L., Araujo-Viel, M. S., Ling, N., and Rossier, J.,** A met-enkephalin-containing-peptide, BAM 22P, as a novel substrate for glandular kallikreins, *Biochem. Biophys. Res. Commun.,* 112, 366, 1983.

166. **Rossier, J., Dean, D. M., Undenfriend, S., and Livett, B. G.,** Enkephalins, congeners, and precursors are synthesized and released by primary cultures of adrenal chromaffin cells, *Life Sci.,* 28, 7181, 1981.

167. **Liston, D. and Rossier, J.,** Synenkephalin is coreleased with met-enkephalin from neuronal terminals in vitro, *Neurosci. Lett.,* 48, 211, 1984.

168. **Patey, G., Cupo, A., Mazarguil, H., Morgat, J.-L., and Rossier, J.,** Release of proenkephalin-derived opioid peptides from rat striatum in vitro and their degradation, *Neuroscience,* in press.

169. **Liston, D., Bohlen, P., and Rossier, J.,** Purification from brain of synenkephalin, the N-terminal fragment of proenkephalin, *J. Neurochem.,* 43, 335, 1984.

170. **Liston, D., Vanderhaegen, J.-J., and Rossier, J.,** Presence in brain of synenkephalin, a proenkephalin-immunoreactive protein which does not contain enkephalin, *Nature (London),* 302, 62, 1983.

171. **Rossier, J., Audiger, Y., Ling, N., Cros, J., and Udenfriend, S.,** Met-enkephalin-Arg^6-Phe^7, present in high amounts in brain of rat, cattle and man is an opioid agonist, *Nature (London),* 288, 88, 1980.

172. **Mizuno, K., Minamino, N., Kangawa, K., and Matsuo, H.,** A new endogenous opioid peptide from bovine adrenal medulla: isolation and amino-acid sequence of a dodecapeptide (bam-12-p), *Biochem. Biophys. Res. Commun.,* 95, 1482, 1980.

173. **Matsuo, H., Miyata, A., and Mizuno, K.,** Novel C-terminal amidated opioid peptide in human phaechromocytoma tumour, *Nature (London),* 305, 721, 1983.

174. **Weber, E., Esch, F. S., Bohlen, P., Paterson, S., Corbett, A. D., McKnicht, A. T., Kosterlitz, H. W., Barchas, J. D., and Evans, C. J.,** Metorphamide: isolation, structure, and biological activity of an amidated opioid octapeptide from bovine brain, *Proc. Natl. Acad. Sci. U.S.A.,* 80, 7362, 1983.

175. **Lang, R. E., Taugner, G., Gaida, W., Ganten, D., Kraft, J. L., Unger, T., and Wunderlich, I.,** evidence against co-storage of enkephalins with noradrenaline in bovine adrenal medullary granules, *Eur. J. Pharmacol.,* 86, 117, 1983.

176. **Marley, P. and Livett, B. G.,** Metorphamide, a novel endogenous adrenal opioid, inhibits nicotine-induced secretion from bovine chromaffin cells, *Neurosci. Lett. Suppl.,* 19, S82, 1985.

177. **Livett, B. G., Marley, P., Mitchelhill, K. I., Wan, D. C.-C., and White, T. D.,** Assessment of adrenal chromaffin cell secretion: presentation of four techniques, in *The Secretory Process,* Vol. 3, Poisner, A. and Trifaró, J. M., Eds., Elsevier/North-Holland, Amsterdam, 1987, 177.

178. **Castanas, E., Bourhim, N., Giraud, P., Boudouresque, F., Cantau, P., and Oliver, C.,** Interaction of opiates with opioid binding sites in the bovine adrenal medulla: cI. Interaction with and μ sites, *J. Neurochem.,* 45, 677, 1985.

179. **Chavkin, C., Cox, B. M., and Goldstein, M. M.,** Stereospecific opiate binding in bovine adrenal medulla, *Mol. Pharmacol.,* 15, 751, 1979.

180. **Costa, E., Guidotti, A., and Saiani, L.,** Opiate receptors and adrenal medullary function (reply), *Nature (London),* 288, 303, 1982.

181. **Costa, E., Guidotti, A., Hanbauer, I., and Saiani, L.,** Modulation of chromaffin cell nicotinic receptors by opiate recognition sites highly selective for met^5-enkephalin[Arg^{6}Phe^7], *Fed. Proc.,* 42, 2946, 1983.

182. **Kumakura, K., Guidotti, A., Yang, H.-Y. T., Saiani, L., and Costa, E.,** Role for the opiate peptides that presumably co-exist with acetylcholine in splanchnic nerves, *Adv. Biochem. Psychopharmacol.,* 22, 571, 1980.

183. **Kumakura, K., Karoum, F., Guidotti, A., and Costa, E.,** Modulation of nicotinic receptors by opiate receptor antagonists in cultured adrenal chromaffin cells, *Nature (London),* 283, 489, 1980.

184. **Saiani, L. and Guidotti, A.,** Opiate receptor-mediated inhibition of catecholamine release in primary cultures of bovine adrenal chromaffin cells, *J. Neurochem.,* 39, 1669, 1982.

185. **Kageyama, H. and Guidotti, A.,** Effect of a-bungarotoxin and etorphine on acetylcholine-evoked release of endogenous and radiolabelled catecholamines from primary cultures of adrenal chromaffin cells, *J. Neurosci. Methods,* 10, 9, 1984.

186. **Castanas, E., Bourhim, N., Giraud, P., Boudouresque, F., Cantau, P., and Oliver, C.,** Interaction of opiates with opioid binding sites in the bovine adrenal medulla. II. Interaction with K sites, *J. Neurochem.,* 45, 688, 1985.

187. **Belcher, G. and Ryall, R. W.,** Substance P and renshaw cells: a new concept of inhibitory synaptic interactions, *J. Physiol. (London),* 272, 105, 1977.

188. **Ryall, R. W. and Belcher, G.,** Substance P selectively blocks nicotinic receptors on renshaw cells: a possible synaptic inhibitory mechanism, *Brain Res.,* 137, 376, 1977.

189. **Ryall, R. W.,** Modulation by substance P of cholinergic transmission, in *Substance P in the Nervous System,* O'Conner, M., Ed., Ciba Foundation Symp. No. 91, Pitman, New York, 1982.

190. **Davies, J. and Dray, A.,** Substance P and opiate receptors, *Nature (London),* 268, 351, 1977.

191. **Krnjevic, K. and Lekic, D.,** Substance P selectively blocks excitation of renshaw cells by acetylcholine, *Can. J. Physiol. Pharmacol.,* 55, 958, 1977.

192. **Steinacker, A. and Highstein, S. M.,** Pre- and post-synaptic action of substance P at the mauthner fibre-giant fiber synapse in the hatchet fish, *Brain Res.,* 114, 128, 1976.

193. **Steinacker, A.,** Calcium-dependent pre-synaptic action of substance P at the frog neuromuscular junction, *Nature (London),* 267, 268, 1977.

194. **Akasu, T., Kojima, M., and Koketsu, K.,** Substance P modulates the sensitivity of the nicotinic receptor in amphibian cholinergic transmission, *Br. J. Pharmacol.,* 89, 123, 1983.

195. **McQueen, D. S.,** Effects of substance P on carotid chemoreceptor activity in the cat, *J. Physiol. (London),* 302, 31, 1980.

196. **Krivoy, W. A., Couch, J. R., Henry, J. L., and Stewart, J. M.,** Synaptic modulation by substance P, *Fed. Proc.,* 38, 2344, 1979.

197. **Kupferman, I.,** Modulatory actions of neurotransmitters, *Ann. Rev. Neurosci.,* 2, 447, 1979.

198. **Mizobe, F., Dean, D. M., and Livett, B. G.,** Peptide modulation of adrenal paraneurons, *Soc. Neurosci. Abstr.,* 5, 534, 1979.

199. **Baker, P. F. and Knight, D. E.,** Calcium control of exocytosis and endocytosis in bovine adrenal medullary cells, *Philos. Trans. R. Soc. London B,* 96, 83, 1981.

200. **Boksa, P., St. Pierre, S., and Livett, B. G.,** Characterization of substance P and somatostatin receptors on adrenal chromaffin cells using structural analogues, *Brain Res.,* 245, 275, 1982.
201. **Kilpatrick, D. L., Slepetis, R., and Kirshner, N.,** Ion channels and membrane potential in stimulus-secretion coupling in adrenal medulla cells, *J. Neurochem.,* 36, 1245, 1981.
202. **Snyder, S. H. and Goodman, R. R.,** Multiple neurotransmitter receptors, *J. Neurochem.,* 35, 5, 1980.
203. **Clapham, D. E. and Neher, E.,** Substance P reduces acetylcholine-induced currents in isolated bovine chromaffin cells, *J. Physiol. (London),* 347, 255, 1984.
204. **Role, L. W.,** Substance P modulation of acetylcholine-induced currents in embryonic chicken sympathetic and ciliary ganglion neurons, *Proc. Natl. Acad. Sci. U.S.A.,* 81, 2024, 1983.
205. **Kilpatrick, D. L., Slepetis, R., and Kirshner, N.,** Inhibition of catecholamine secretion from adrenal medulla cells by neurotoxins and cholinergic antagonists, *J. Neurochem.,* 37, 125, 1981.
206. **Marier, S. H., Putney, J. W., Jr., and Van De Waller, C. M.,** Control of calcium channels by membrane receptors in the rat parotid gland, *J. Physiol. (London),* 279, 141, 1978.
207. **Hanley, M. R., Sandberg, B. E. B., Lee, C. M., Iversen, L. L., Brundishy, C. E., and Wade, R.,** Specific binding of ^{3}H-substance P to rat brain membranes, *Nature (London),* 286, 810, 1980.
208. **Rink, T. J.,** The influence of sodium on calcium movements and catecholamine release in thin slices of bovine adrenal medulla, *J. Physiol. (London),* 226, 297, 1977.
209. **Tregear, G. W., Niall, H. D., Potts, J. T., Leeman, S. E., and Chang, M. M.,** *Nature (London), New Biol.,* 232, 87, 1971.
210. **Quirion, R.,** Multiple tachykinin receptors, *TINS,* May, 183, 1985.
211. **Boksa, P. and Livett, B. G.,** The substance P receptor subtype modulating catecholamine release from adrenal chromaffin cells, *Brain Res.,* 332, 29, 1985.
212. **Erjavec, F., Lembeck, F., Florjanc-Irman, H., Skofitsch, G., Donnerer, J., Saria, A., and Holzer, P.,** Release of histamine by substance P, *Naunyn-Schmeideberg's Arch. Pharmacol.,* 317, 67, 1981.
213. **Johnson, A. R. and Erdos, E. G.,** Release of histamine from mast cells by vasoactive peptides, *Proc. Soc. Exp. Biol. Med.,* 142, 1252, 1973.
214. **Sydbom, A.,** Histamine release from isolated rat mast cells by neurotensin and other peptides, *Agents Actions,* 12, 91, 1982.
215. **Araki, T., Hamamoto, T., Kurosawa, M., and Sato, A.,** Response of adrenal efferent nerve activity to noxious stimulation of the skin, *Neurosci. Lett.,* 17, 131, 1980.
216. **Araki, T., Ito, K., Kurosawa, M., and Sato, A.,** The somato-adrenal medullary reflexes in rat, *J. Auton. Nerv. Syst.,* 3, 161, 1981.
217. **Lembeck, F. and Skofitsch, G.,** Visceral pain reflex after pretreatment with capsaicin and morphine, *Naunyn-Schmiederberg's Arch. Pharmacol.,* 321, 116, 1982.
218. **Haeussler, G. and Osterwalder, R.,** Evidence suggesting a transmitter or neuromodulatory role for substance P at the first synapse of the baroreceptor reflex, *Naunyn-Schmiedeberg's Arch. Pharmacol.,* 314, 111, 1980.
219. **Niijima, A. and Winter, D. L.,** Baroreceptor in adrenal gland, *Science,* 159, 434, 1968.
220. **Livett, B. G., Kozousek, V., Mizobe, F., and Dean, D. M.,** Substance P inhibits nicotinic activation of chromaffin cells, *Nature (London),* 278, 256, 1979.
221. **Mizobe, F., Kozousek, V., Dean, D. M., and Livett, B. G.,** Pharmacological characterization of adrenal paraneurons: substance P and somatostatin as inhibitory modulators of the nicotinic response, *Brain Res.,* 178, 555, 1979.
222. **Mizobe, F., Dean, D. M., Rorstad, O., Whelan, L. G., and Livett, B. G.,** Functional studies on adrenal paraneurons: substance P and somatostatin as inhibitory neuromodulators of adrenergic cells, in *Neurotoxins,* Union Press, Adelaide, Australia, 1979, 267.
223. **Tatemoto, K. and Mutt, V.,** Isolation of two novel candidate hormones using a chemical method for finding naturally occurring polypeptides, *Nature (London),* 285, 417, 1980.
224. **Tatemoto, K.,** Isolation and characterisation of peptide YY (PYY), a candidate gut hormone that inhibits pancreatic exocrine secretion, *Proc. Natl. Acad. Sci. U.S.A.,* 79, 2514, 1982.
225. **Adrian, T. E., Terenghi, G., Brown, M. J., Allen, J. M., Bacarese-Hamilton, A. J., and Pollak, J. M.,** *Lancet,* 2, 540, 1980.
226. **Lundberg, J. M., Terenius, L., Hokfelt, T., and Goldstein, M.,** High levels of neuropeptides Y in peripheral noradrenergic neurons in various mammals including man, *Neurosci. Lett.,* 42, 167, 1983.
227. **Allen, J. M., Adrian, T. E., Pollak, J. M., and Bloom, S. R.,** Neuropeptide Y (NPY) in the adrenal gland, *J. Auton. Nerv. Syst.,* 9, 559, 1984.
228. **Lundberg, J., Terenius, L., Hökfelt, T., Matling, C. R., Tatemoto, K., Mutt, V., Polak, J., Bloom, S., and Goldstein, M.,** Neuropeptide Y (NYP)-like immunoreactivity in peripheral noradrenergic neurons and effects of NPY on sympathetic function, *Acta Physiol. Scand.,* 116, 477, 1982.
229. **Ekblad, E., Edvinsson, L., Wahlestedt, C., Uddman, R., Hakanson, R., and Sundler, F.,** Neuropeptide Y co-exists and cooperates with noradrenaline in perivascular nerve fibres, *Reg. Peptides,* 8, 225, 1984.

230. **Varndell, I. M., Polak, J. M., Allen, J. M., Terenghi, G., and Bloom, S. R.,** Neuropeptide tyrosine (NPY) immunoreactivity in norepinephrine-containing cells and nerves of the mammalian adrenal gland, *Endocrinology,* 114, 1460, 1984.

231. **Maebashi, M., Mouri, T., Itoi, K., and Yoshinaga, K.,** Elevation of blood pressure in rats by passive immunization against neuropeptide Y, *Nature (London),* submitted.

232. **Khalil, Z., Marley, P. D., and Livett, B. G.,** Insulin-stress-induced catecholamine secretion from the adrenal gland in-vivo: neurogenic and non-neurogenic components, *Neurosci. Lett.,* 19(Suppl.), S78, 1985.

233. **Khalil, Z., Marley, P. D., and Livett, B. G.,** The pattern of adrenal catecholamine secretion in response to insulin-induced hypoglycaemia is related to the mechanism of adrenal stimulation, *Proc. Aust. Endocrinol. Soc.,* 28, 93, 1985b.

234. **Rosenfeld, M. G., Mermod, J.-J., Amara, S. G., Swanson, L. W., Sawchenko, P. E., Rivier, J., Vale, W. W., and Evans, R. M.,** Production of a novel neuropeptide encoded by the calcitonin gene via tissue-specific RNA processing, *Nature (London),* 304, 129, 1983.

235. **Wan, D. C.-C. and Livett, B. G.,** unpublished.

236. **Dean, D. M., Livett, B. G., and Leeman, S. E.,** unpublished results.

237. **Mitchelhill, K. I., Marley, P., and Livett, B. G.,** unpublished.

238. **Khalil, Z., Marley, P., and Livett, B. G.,** unpublished results.

239. **Marley, P., Oliver, C., and Livett, B. G.,** unpublished.

INDEX

A

A23187, see Ionophore A23187
Acetylcholine
  action potential and, 76, 77
  adrenergic response and, 126
  calcium and, 60
  calcium flux and, 83, 123
  cAMP and, 53
  cGMP and, 54, 55
  concentration of, 122
  desensitization and, 103
  dynorphins and, 129
  insulin stress and, 126
  membrane capacitance and, 89, 90
  membrane depolarization by, 75
  muscarinic receptors and, 63—65, 125—126
  opioid peptides and, 131
  phosphoinositide turnover and concentration of,
    55
  in salivary gland, 120
  splanchnic nerves and, 118
  substance P and, 136
Acetylcholine channels, 92—94
Acetylcholine-dependent calcium channels, 88
Acetylcholine-induced potential fluctuation, 118
Acetylcholine receptors, 91—92
Acetylcholinesterase, 46
Acinar cells, 136
Aconitine, 72—74
ACTH, 121
Actin, 5, 6, 39, 41
F-Actin, 5, 6, 29
G-Actin, 28
α-Actinin, 5
Action potentials, 76, 77, 119
Adenosine monophosphate (AMP), cyclic, see
    Cyclic AMP
Adenosine triphosphatase (ATPase), 6, 38
Adenosine triphosphate (ATP), 2, 3, 30, 46, 126
Adenylate cyclase, 38
Adrenal cell cultures, 122
Adrenal medulla
  development of, 120
  electrophysiological properties of cells of, 75—77
  liposome fusion and, 24
  neuropeptides in, 127—130
  phosphoinositide turnover in, 55—57
  synexin in, 6
Adrenergic receptor responses, 126—127
Adrenocorticosteroids, 65
Adrenorphin (metorphamide), 121, 130, 132
Alpha adrenergic blockers, see also specific types,
    27
Amantidine, 136
Amidorphin, 133
Amiloride, 29
Anesthesia, 120, 136

Angiotensin, 75
Angiotensin II, 140
Anoxia, 120
Anticalmodulin drugs, see also specific types, 5
Anticytoskeletal drugs, see also specific types, 5, 27
Antisynexin drugs, 5
Arachidonic acid, 7, 9, 42, 58—59
Assays, see also specific types
  of cell-liposome interactions, 17—19
ATP, see Adenosine triphosphate
Atropine, 121
  calcium and, 60
  membrane depolarization and, 75
  muscarinic receptors and, 63, 65
  phosphoinositide turnover and, 55
  plasma membranes and, 38

B

Bacitracin, 132
BAM-22P, 121, 129
BAM peptides, see also specific types, 132
Barium
  calcium and, 102, 106—107
  calcium flux and, 82
  desensitization and, 124
  membrane depolarization by, 75
  metalloendoprotease inhibitors and, 4
  in sarcoplasmic reticulum, 107
Basophilic leukemia cells, 19
Batrachotoxin (BTX), 72—74, 77, 84
BAY-K-8644, 99, 102—104, 107—108, 111
Bestatin, 132, 133
Beta adrenergic receptors, see also specific types,
    65
BGTX, see α-Bungarotoxin
Big unitary conductance calcium-dependent potas-
    sium channels (BK channels), 119
Bioluminescence, see Chemiluminescence
BK channels, see Big unitary conductance calcium-
    dependent potassium channels
Blood pressure, 138
Bovine pancreatic polypeptide (BPP), 140
Bovine serum albumin (BSA), 20, 21
BPP, see Bovine pancreatic polypeptide
Bradykinin, 75, 140
Brain synaptosomes, 81
p-Bromophenacyl bromide, 58
BSA, see Bovine serum albumin
BTX, see Batrachotoxin
α-Bungarotoxin, 53, 76, 122, 124
BW 755-c, 58

C

Cadmium, 81